D0455529

INTRODUCTION TO THE

INTRODUCTION TO THE GLOBAL OIL & GAS BUSINESS

SAMUEL A. VAN VACTOR
PennWell®

PennWell Corporation
1421 South Sheridan Road
Tulsa, Oklahoma 74112-6600 USA

800.752.9764
+1.918.831.9421
sales@pennwell.com
www.pennwellbooks.com
www.pennwell.com

Marketing: Jane Green
National Account Executive: Barbara McGee

Director: Mary McGee
Managing Editor: Stephen Hill
Production Manager: Sheila Brock
Production Editor: Tony Quinn
Book Designer: Susan E. Ormston
Cover Designer: Alan McCuller

Library of Congress Cataloging-in-Publication Data

Van Vactor, Samuel A.
Introduction to the global oil and gas business / Samuel Van Vactor.
p. cm.
Includes bibliographical references and index.
ISBN 978-1-59370-214-4
1. Petroleum industry and trade. 2. Gas industry. I. Title.

HD9560.5.V36 2009
338.2'728--dc22

2009045762

Printed in the United States of America

5 6 7 8 9 19 18 17 16 15

CONTENTS

ACKNOWLEDGMENTS

I wish to thank Arianna Lambie, Fredrick Pickel, David Ramberg, Valerie Karplus, Robert Mullen, Michael Telson, and Arlon Tussing for helpful comments. Any errors or omissions are solely the responsibility of the author.

1

INTRODUCTION

The Demand and Supply of Oil

At the close of Jimmy Carter's presidency, a reporter asked his chief of staff, Hamilton Jordan, if he had any advice for future presidents. Jordan had a straightforward observation: avoid running for reelection in an energy crisis.

Hardly anything stirs public anger like runaway gasoline prices. The widespread availability of oil seems an essential part of modern life, but it was not always so; oil did not become the backbone of the global economy until the second half of the twentieth century. Coal used to be far more important. Oil became the dominant form of energy following a massive wave of Middle East discoveries just before and after World War II. The industry found so much oil that at times they could not give it away. Today the circumstances have reversed. Some claim that the modern world's dependence on oil corresponds to that of an addict—no one can get sober enough, long enough, to break the habit. At the market's peak in 2008, *daily* crude oil production had a cash value of $13 billion, a sum larger than the annual turnover of many industries. So, when prices are on the upswing, all the ensuing uproar raises a key question: Why is oil so valuable?

Much of the answer lies in the unique qualities that make oil a particularly advantageous source of energy. Once found, oil is the easiest of all energy commodities to produce and use. There is little waste in producing and converting crude oil into useful products. Since oil is liquid, it moves cheaply by a variety of methods. As a liquid, oil converts readily into a gas in order to power an engine. Compared to coal, it is less bulky and much cleaner. These advantages are especially valuable for use in transportation, which is oil's primary market.

Arguably, oil is the most efficient form of energy, but it requires vast long-term investments to produce and use. It is the second part of this relationship that raises most political and social issues. In order to use oil, consumers must make large fixed-cost investments in specialized equipment— cars, trucks, furnaces, and so on. Those investments depend on a reliable flow of petroleum products. If there are shortages, or if prices explode, consumers feel cheated. Inflexible investments in oil-producing and in using equipment affect the market in another way. The investments lock everyone in, because when prices go up, it is hard to produce more quickly or cut back consumption. Thus, the price mechanism is not always effective in balancing demand and supply. This leads to great price volatility.

Oil supplies are concentrated in a few key countries, mostly in remote parts of the globe. In contrast, oil consumption takes place mainly in developed countries in North America, Europe, and the coastal regions of Northeast Asia. Unfortunately, there is little overlap and the parties have widely divergent interests. Historically, this has exaggerated motives to colonize or otherwise control oil flows, which, in turn, has bred suspicion among oil-exporting countries. Anxiety over oil supplies has also led to conflicts among consuming nations over access and pricing.

The oil shocks of the 1970s led many, including a U.S. president, to expect permanent shortages. Instead, oil consumers enjoyed a surplus during the two decades that followed. The tables turned with the new century. Geologists warned of an eminent peak in oil supply, claiming that the era of great discoveries was over. Many of them put the peak at around 2005, after which oil production would decline, no matter what producers did. It is still unproven as to whether or not the "peak oil" advocates are right or wrong, but in 2008, extraordinary economic growth strained production capacity, and oil prices surged to record levels.

The great flaw in the peak oil argument is the exclusive focus on conventional oil supplies—oil that is the easiest to find and cheapest to produce. The resource base of unconventional hydrocarbons in the earth's crust is staggering, many times the present estimate of oil and gas reserves, with a lifetime of centuries not years. This supply diversity creates a staircase of increasing cost, and as producers mount new stairs, technology improves and costs came down. Although the portfolio of unconventional oil supply is vast, there is a fly in the ointment. Processing conventional crude oil to useful products is cheap and energy efficient because there is relatively little waste and environmental impact. In contrast, processing most unconventional hydrocarbon resources into useful products requires a great deal of energy in and of itself and often produces undesirable side effects. For example, it takes about one-third of the thermal content of a barrel of oil to process oil sands into a useable barrel of product. Similarly, producing ethanol from sugar or corn requires a great deal of energy. Some scientists argue that it takes more energy to produce corn ethanol than can be extracted from the final product.

In a market-based system, energy consumers are unlikely to face shortages, but if the peak oil enthusiasts are right, those consumers will instead see higher prices driven by increased costs and environmental constraints. For many, this amounts to the same thing as a shortage.

Oil Prices

Even Agatha Christie could not invent a better mystery than that which surrounds oil prices, particularly when they zoom upward. The suspects abound—speculators, OPEC, and the Chinese are recent persons of interest. Apparently, the public has cleared major oil companies of any wrongdoing, but they were prime suspects in the 1970s. Of course, the interaction of demand and supply sets oil prices. That, however, is cold comfort to consumers that depend on petroleum products at reasonable prices, and it is an unenlightening catch phrase.

There are two main theories about how oil prices are determined. The traditional view, offered by Harold Hotelling, sees oil as a depleting resource fixed in quantity. Given an absolute limit, the oil may be consumed slowly or quickly, but once gone, it's gone. Prices reflect demand over a series of years until supply is exhausted. The allocation of oil between current and future periods depends on annual demand, total resources, production costs, alternative energy costs, and the interest rate. If the current price is too low, producers will shut down wells and sell the oil later when prices approach alternative costs. If oil prices move too high, producers will ramp up production rates to take advantage of a better return on capital, bringing prices back to earth. The dynamic interplay of these key variables determines what everyone pays at the gasoline pump.

Alternatively, Morris Adelman dismisses the notion that oil supplies are a fixed quantity. He argues that current production rates and the level of reserves that support them are flexible; they depend on current technology and the incentive to explore. If oil prices rise, drillers get off their duff and explore. If prices go down, exploration funds dry up until the market eventually turns around. If reserves become harder to find, oil prices will rise until entrepreneurs invent substitutes. According to Adelman, the variables that determine prices in the Hotelling model are largely "unknown and unknowable" and have little bearing on daily oil prices.

The oil market, like many commodity markets, concentrates trading on a benchmark or "marker." Benchmarks are important for trading, because crude oil varies in quality and each type of crude oil has a separate price. A benchmark establishes the general trend. In today's market, there are two key benchmarks: light sweet crude oil trading in the U.S. Midwest and North Sea oil trading in the United Kingdom. These oils became benchmarks because the New York Mercantile Exchange (NYMEX) and Intercontinental Exchange (ICE) chose them as the basis for futures trading.

The OPEC cartel fixed crude oil prices directly for over a decade, but abandoned the effort in 1986 when the market collapsed. Today OPEC manages the market indirectly by establishing production quotas for its members.

Cartel members have underinvested in resource development for decades, and, consequently, they did not have enough spare capacity to meet a surge in demand during 2008. The upshot was runaway prices followed by a market collapse.

Oil exchanges have existed at various times in the history of the industry, but none survived until NYMEX began trading heating oil in 1978. Traditionally, major companies "posted" prices for various crude oils they were willing to buy or sell. The postings allowed flexibility, as they could change with market conditions, but most of the time they were stable. A small "spot" market—about 5% of the total—supplemented the posting system. It acted as a balance wheel, allowing refiners to adjust processing runs as conditions changed.

Oil traders have experienced several types of price indexing. In 1985, Saudi Arabia indexed its crude oil prices to a "netback" of petroleum product prices plus a refinery margin. The purpose of the scheme was to preserve Saudi market share in a deteriorating market. A far more common use of indexing concerns oil purchase contracts. Survival in the oil business requires flexibility, particularly in the face of volatile prices. To meet the challenge, traders tie contract prices to those published in the trade press or made available from the futures exchanges, rather than setting a flat price that might be unrealistic.

There are seven primary regions for oil trading— the North Sea, the Russian-Caspian region, the Mediterranean, West Africa, the Mideast Gulf, Asia, and the Americas. Since the introduction of super tankers, crude oil can be shipped almost anywhere in the globe for $2 per barrel or less. The resulting fungibility of the market causes oil prices to move up and down in tandem. Nonetheless, there are regional differences, with variation in crude oil quality, transportation options, and different institutional arrangements. The trade press tracks the key markets and reports prices on all the important crude oils.

Commodity markets today are a complex mix of physical and financial trading. Physical trading, as the name implies, is the purchase and delivery of the actual commodity. In the case of crude oil, it sells mainly by tanker cargo or by daily flows on a pipeline.

Energy Commodity Markets

Crude oil futures markets trade huge volumes, roughly eight times the global physical flow of oil. The primary purpose of the trade is to manage price risk and set the general trend, not to move oil from fields to refineries. Futures contracts are derivatives—that is, the contracts derive their value from the value of an underlying asset. Futures traders do not trade the commodity, they trade the right to buy or sell the commodity. Indeed, the NYMEX contract identifies only one type of crude oil at one location—light sweet oil at Cushing, Oklahoma. Even though futures traders price a derivative contract, it corresponds directly to prices in the physical market and affects all oil transactions.

Futures exchanges provide important benefits. In today's market, they are the primary means of price discovery. That is, they are the forum in which those that wish to buy oil interact with those willing to sell. Many markets take place behind closed doors, and neither the activity nor the outcome is visible. In contrast, futures markets are entirely transparent. Buyers (or sellers) may not like the result, but they can watch the process. The high volume of trading in oil futures markets creates liquidity. Anyone can buy or sell oil with the click of a computer button. This makes oil consumers more secure, since oil will always be available at the market price. As a formal exchange, futures markets reduce counterparty risk. If a trader's position deteriorates, the exchange closes the contract before default can occur. Futures trading, or any activity concerning derivatives, shifts risk to those better prepared to accept it. Using futures markets, refiners, industrial oil consumers, or anyone can lock in prices or rates of return, ignoring price volatility.

Not all futures contracts succeed. In fact, most fail. For a contract to succeed it must have the following characteristics: 1) there must be price uncertainty; 2) the product may be heterogeneous, but there must be correlation between the various grades or locations; 3) there must be a large number of active traders; 4) the product has to have significant value; and 5) the market must be unregulated. When oil trading started at NYMEX, federal authorities were phasing out price controls and other types of regulation. President Reagan's first act as president in January 1981 was to terminate the oil regulatory program. The timing was brilliant, because oil prices had peaked and would spiral downward for five more years making deregulation a big success.

Futures exchanges trade contracts for each month going forward into the future, as well as options—the right to buy or sell oil in the future at a given price level. Contracts range out about eight years. This price information is an important signal for oil producers, helping to rationalize investment decisions. When prices in far months are below prices in near months, the industry describes the term structure as *backwardation*. In reverse, when far-forward prices are higher, the descriptive term is *contango*. Commonly, the market is in contango, but given strong immediate demand, it can switch to backwardation. This happens in emergencies, such as hurricanes or other short-term disruptions.

Futures markets work in tandem with over-the-counter (OTC) trading. While the futures market sets the general trend for prices, and OTC trading works out relative prices for different quality crude oils located all over the globe. The two types of trade are essential complements. Without reliable relative prices, traders could not use futures contracts to hedge. Likewise, without a liquid and transparent trade in the exchanges' benchmark prices, the OTC would be far less efficient.

The New Industry Structure

The oil industry is cyclical, despite persistent alarms of imminent and permanent shortages. It expands or contracts primarily in response to the short-term abundance or scarcity of crude oil. Scarcity normally provokes rising prices and the industry responds with a frantic expansion of exploration and development. Then, exploration success follows success, demand softens, prices drop, and the industry shrinks. The cycle's rhythm is hard to pin down, but measures in decades instead of years. Each generation must learn anew the opportunities and perils of growth and contraction.

In most of the globe, governments are reluctant to step back and allow oil companies carte blanche in their operations. Government officials in countries dependent on foreign suppliers often view oil as a strategic necessity and seek measures to ensure its supply. The leaders of oil-exporting countries fear the unfettered exploitation of their resources and, more cynically, recognize all too clearly the great wealth that prolific oil fields can bestow.

Originally, the interests of consumers trumped the opportunities of suppliers. In the twentieth century, national governments with strategic interests in oil supplies sponsored or otherwise controlled many of today's great oil corporations—BP from the UK, Total from France, and ENI from Italy are just a few examples. In contrast, the educated elite from oil exporting countries discovered to their horror an earlier phrase from nineteenth century colonialism: "For god's sake, don't tell the natives what it's worth." It took decades, however, before the oil exporters could assert control, and once they did, the oil world changed forever.

The currents of public and private interests swirl through today's industry. Essentially, there are four types of oil companies: First, are traditional companies, international oil companies (IOCs,) that operate mainly from consuming nations and often have a century-long history. Private investors own these companies and government activity is limited to regulation and oversight policies. IOCs operate all over the globe, taking advantage of investment opportunities or emerging markets.

Second, are the national oil companies (NOCs.) Governments, mostly from OPEC nations, own and control these companies. Many of the NOCs began life as government ministries, but as the IOCs withdrew, NOCs replaced them in daily operations and investment activities. In other cases, governments bought out the operating company or companies of the relevant IOCs, and integrated local personnel into the new company. With rare exceptions, the NOCs stay close to home and simply manage local resources, often providing the prime source of government revenue.

As oil prices began to rise after the low point in 1998, a new class of company gained popularity. China's Petro China typifies these companies, in which the government owns the majority interest, but allows private investors to buy shares. These companies may be categorized as government sponsored enterprises (GSEs) after the acronym devised to describe U.S. mortgage companies Fannie Mae and Freddie Mac.

Although the trend is recent, the idea is not new, going all the way back to the British government's investment in BP during World War I. The present environment has yet to test success or failure of this organizational structure. Historically, government sponsorship has waned during weak markets, as the pain of subsidies grows.

With a few exceptions, OPEC governments have locked out IOCs from the most prolific oil-producing regions. This has one largely overlooked consequence—a reduction in size and number of employees among the IOCs and a rapid growth of the oil service sector. Slumberger's annual revenue in 2007 was $233 billion, and only three of the IOCs were higher. OPEC members are willing to contract with Western service companies, but remain shy about dealings with IOCs.

Low oil prices in the late 1990s drove a major consolidation of IOCs. As they shrunk, GSEs and service companies expanded. Subtle shifts over the years are breaking down the historical IOC structure. Vertical integration has eroded as more and more companies specialized, focusing on their most profitable sectors. In contrast, GSE's seek to retain an integrated structure. Companies like Petro China, OMV of Austria, and ONGC of India are actively developing resources abroad in order to balance crude oil flows with their refinery capacity.

Energy Security

Given that efficient commodity markets exist for crude oil supplies, why is energy security an issue? This is a question seldom asked, even as most oil-consuming countries subsidize massive investments in alternative energy.

Part of the answer is a matter of history. Access to oil supplies decided the outcome of many World War II battles and provoked conflicts throughout the industry's history. Most of these issues belong to the past; militaries currently face small regional conflict or a short all-out war. A residual of unease remains, however, and policies preach the benefits of energy independence. So far, policies aimed at accelerating the development of alternative energy supplies have not made much of a dent. In the last two decades of the Twentieth Century, U.S. oil dependence doubled, despite the efforts of both Presidents Nixon and Carter to encourage conservation and domestic alternatives.

Government officials from most countries dependent on foreign oil suppliers perceive that their economies are at risk from high oil prices, even if their consumers can still buy fuel. Individual consumers often feel the same way, stuck, as they are, with furnaces, cars, trucks, and tractors that depend on reasonably priced fuel.

It has been more than three decades since Arab OPEC, dissatisfied with pro-Israeli policies in the Netherlands and the U.S., triggered the "oil weapon." Since oil is fungible, the weapon was easily defused. Even an ineffective weapon, however, may shift political attitudes. While the U.S. did not abandon Israel, it did awake to growing power of the Arab oil-producing states and that is now a foreign policy reality.

Organizing Principals

At times, firms allocate resources by markets, and other times, by internal transfer. Economist Ronald Coase pondered this question and concluded that when the transaction costs of using a market are too high, firms expand, encompassing a variety of activities. As market efficiencies improve, firms narrow their focus and specialize. If markets are inefficient then firms' incentives are to integrate vertically—up and down the production chain, controlling the entire process, from raw resources to final products.

In contrast, Paul Frankel argued that the nature of petroleum itself shaped the industry's structure. The central idea followed from the observation of early twentieth century architects—*form follows function*. Frankel tied his ideas together with traditional economic thought; that is, economies of scale in transporting oil lead to market concentration and, in his view, vertical integration. Frankel's ideas were popular with major oil companies, but frequently criticized by mainstream economists, who regarded the argument as a self-serving rationalization for undeserved market power.

The oil industry's unusual characteristics set it apart from commonplace manufacturing. As Frankel noted, oil is in liquid form and is easily combustible, thus, it requires special handling. Industry old-timers also describe the process as "continuous flow." Oil flows from well to refinery to consumers, without slowing down. If the flow is disturbed, as it was during Hurricanes Katrina and Rita, shortages show up quickly. The oil industry is one of continuous flow, but it is not unique—electricity, natural gas, transportation services, sports, and theatre all depend on schedules and timing.

OPEC's revolution, and its aftermath, demonstrates the validity of Coase's approach. Justified or not, in 1973, OPEC severed the oil industry's industrial structure. The cartel attempted to replace the internal transfers of major oil companies with long-term contracts based on each country's official prices. The system worked for about five years, but the Iranian Revolution and a runaway spot market tore it apart. Many OPEC members abrogated purchase contracts with little or no warning, leaving buyers stranded. Incomplete and opaque price information compounded the uncertainty. In Coase's terms, the transactions costs of buying crude oil increased significantly.

An inefficient market is a burden for some and an opportunity for others. In this case, NYMEX stepped into the breach offering both risk management and reliable price information. Not accidentally, the industry gravitated to futures trading and thereby undermined the OPEC pricing strategy. As the new market structure matured, the transaction costs of trading oil declined.

Before the expansion of the spot oil market and development of futures trading, the goal of most large oil companies was to achieve a balance of oil production, refining and marketing. The motive to integrate was strongest in the international market, where contract enforcement was weakest. While this made each company more secure, it reduced managerial efficiency due to the diversity of skills required. Once it became easy and less costly to buy and trade crude oil, managers in the various segments of the industry could focus on their primary tasks. Put succinctly, when markets work smoothly, companies improve profitability by specializing.

Today, the international oil market is mature, with remarkably low trading costs. Crude oil produced almost anywhere can be sold almost anywhere so the advantages of vertical integration have largely vanished. In contrast, natural gas companies in the global market frequently encounter what economists identify as the "specific asset problem." Energy investments are capital intensive and usually have little or no alternative value, a classic example of sunk costs. In advance of construction, there is significant competition. At any one time, dozens of qualified firms might compete for a particular project. Likewise, there is always a large inventory of potential ventures. Once a fixed capital project is complete, however, competition narrows to just a few parties. For example, if a gas producer develops a field and builds a pipeline to a distribution utility, that producer depends on the buyer fulfilling contractual obligations. If the buyer reneges, a producer with an unenforceable contract may have no choice but to reduce price.

International enforcement is crucial, because cross-border energy projects are now the primary means to link the most prolific resources to the best markets. Since World War II, contract enforcement has improved as a series of treaties and multilateral organizations strengthened the ability of an injured party to redress an injustice. Likewise, the industry has developed complex investment schemes that balance interests, making it difficult for participants to default on their obligations. For example, such schemes often require all parties to invest capital, thus aligning interests.

Alternatives to Oil

Scientists have thought about the shift from fossil fuels to sustainable alternatives for decades. It was the undercurrent of President Eisenhower's *Atoms for Peace* program of the 1950s. Still, the visions then and the visions now are very different. The forthcoming energy transition is nothing less than a marvelous challenge for the twenty-first century. Imperfect as any glimpse of the future is, it reveals a marvel of tantalizing technologies emerging from laboratories and factories around the world.

Different futurists today see vastly different energy futures. The neo-Malthusians view the downward trend of the peak oil thesis as a symbol of economic and social collapse, an apocalyptic future in a world without oil. Others are far more optimistic, visualizing a future in which humans finally seek to live with, rather than, conquer nature.

There are two quite different examples of changing fuel use that demonstrate how energy transitions take place. In the nineteenth century, the primary lamp fuel was whale oil. As demand grew, whalers spread from the Arctic to the Antarctic. Increasing scarcity resulted in smaller harvests and rising prices. This, in turn, provoked an alternative—kerosene, distilled from crude oil. This transition was straightforward and simple, because the delivery and use infrastructure did not have to change. Existing lamps used kerosene directly, without alteration.

The transition from coal to oil as the primary form of energy was far more complex. Since the form and properties of the two fuels are radically different, the entire infrastructure of production, transportation, and use changed. Fundamentally, oil use is entirely different from coal. Oil ignites easily in an internal combustion engine while coal does not. Among other things, this shifted the entire scale and scope of transportation; it became economical to develop small vehicles for use on roads and farms, rather than relying on large steam engines.

The nineteenth-century British academic, William Stanley Jevons, argued that rapid improvements in steam engine design and coal use were the moving forces behind the industrial revolution. This observation rings true today. If lighter materials, fewer moving parts, and less energy achieve the same result, productivity increases, which ultimately means a higher standard of living. This is the essence of the coming transition from oil. Many of the emerging technologies are far more efficient than existing ones. They may cost more at first, but history teaches that, in the long-term, they will cost less.

Engineers typically calculate the cost effectiveness of new energy sources or technologies without considering the broad impact of the choice. There are many issues to consider. Is the primary resource base adequate? For example, it is relatively cheap to convert cars to run on natural gas, but gas reserves may be insufficient. Direct costs, of course, are always a primary consideration. Some choices like synthetic fuels require little or no change in the use infrastructure. This gives them an initial advantage, but high cost and environmental constraints may undo the effort. Other energy sources, like fusion, are elegant in theory, but unproven. Efficient storage is always a critical part of energy use and has been oil's biggest advantage. Storage options for natural gas and electricity are improving, but still fall short of liquid fuels. Environmental side effects and the release of greenhouse gases are huge constraints on options like coal. Although not well defined, energy security remains another factor governing choice.

Oil's primary use is in transportation, and the auto industry is engaged in developing alternatives. So far, electric cars appear to be the most likely alternative, due to recent advances in battery technology and the success of hybrid vehicles. Nonetheless, the various options require review with respect to the weight and volume of the energy source and the overall "well-to-wheel" efficiency of the design. Using these criteria, electric motors, with power from batteries, onboard generators, or fuel cells, have a significant energy efficiency advantage over the internal combustion engine.

Over the next several years, the auto industry plans to produce and market a wide variety of hybrid, plug-in hybrid, and electric vehicles. The speed at which these new types of designs penetrate the market will depend on oil prices. These vehicles cost more than conventional designs, but have much lower running costs. At today's costs, it will take crude oil prices in excess of $100 per barrel to make it worthwhile for consumers to invest in these higher-cost vehicles. This may not seem like a remarkable advance from a decade ago, but it is. For the first time, there are realistic alternatives to oil. They are not immediate, but they do not have to be, since conventional oil supplies will extend for decades.

According to the peak oil advocates, the globe has consumed about one-half of all conventional oil supplies in 151 years. It is likely that the oil industry will be active throughout the twenty-first century under any scenario. Ultimately, oil use will be constrained to activities for which there are few substitutes— jet fuel, heavy trucking, and so forth.

Literally millions of adjustments in consumption and production are required for an efficient transition from oil to alternatives, adjustments far beyond the imagination of the most creative bureaucrats. Shifting relative prices guided the transition from whale oil to kerosene and from coal to petroleum. Likewise, the price mechanism must mandate the reallocation from oil to alternatives if the transition is to be efficient.

Even as conventional oil supplies decline and prices rise, there is no reason to believe that it signals the end to economic growth and prosperity. If Professor Jevons was right—if increasing energy efficiency is the motor of economic development—then oil's great contribution was to underpin the extraordinary global economic growth since World War II. The growth in wealth and economic prosperity has opened the portals of human opportunity. Today, there are vastly more educated people than ever before. This storehouse of knowledge and creative activity will surely discover viable alternatives to depleted oil.

2 THE DEMAND AND SUPPLY OF OIL

Why Is Oil So Valuable?

In July 2008, the trading pit of the NYMEX futures exchange came alive as excited traders bid the price of a barrel of crude oil up to $147 per barrel, the highest price so far recorded. The frenzy climaxed a long upward march in the price of oil. At times in 1998, the commodity had traded for under $11 per barrel. In the space of a decade, oil producers enjoyed a fourteen-fold price increase (fig. 2–1). On that frantic day, it seems no one could do without the world's most sought-after commodity. The recent panic is different in form, but no less shrill than earlier experiences. In World War I, Winston Churchill, Lord of the Admiralty, nationalized Anglo Persian Oil (now BP) to ensure adequate supplies for the Royal Navy. Similarly, a primary motive for the Japanese attack on Pearl Harbor was the U.S. blockade of Indonesian oil supplies. Most people think oil is essential to a modern industrial economy—the one necessity that everyone needs. If so, what makes oil so valuable and why is its price so volatile?

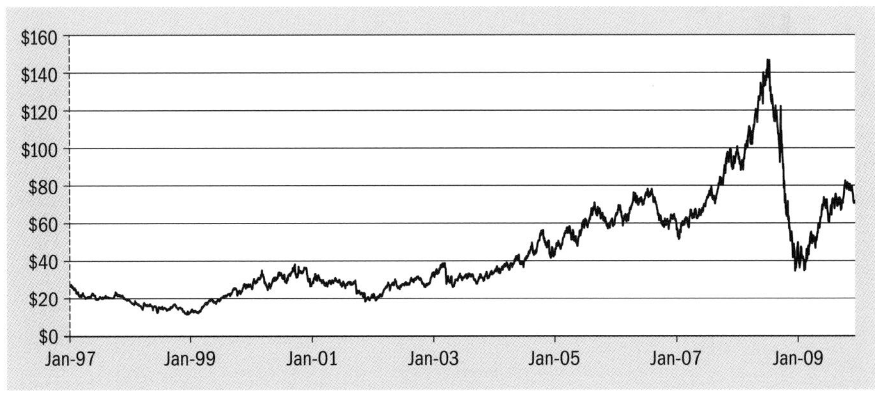

Fig. 2–1. NYMEX prompt crude oil price 1998–2008

Of all energy sources, oil is the most versatile and cheapest to handle and use. Wood and coal are bulky to carry about and dirty when burned. Wires must connect an electric generator to a consumer and, so far, storage is prohibitively expensive. Pipelines deliver natural gas; alternative delivery methods exist, but they are costly. The efficient delivery of both gas and electricity requires a grid—a large fixed cost infrastructure that has little or no alternative use and limits the size of the market. In contrast, oil is flexible. Ocean tankers, pipelines rail cars, and trucks can all deliver oil. Petroleum products divide easily into small lots, and relative to many alternatives, they are safe to use. Most importantly, oil has a high energy density; it packs a wallop for its size and weight. Oil's characteristics make it an ideal transportation fuel and that sector is now its primary market.

Writing a half century ago, Paul Frankel noted the unusual properties of oil and the characteristics that set it aside from other energy sources and other commodities. He noted that kerosene and fuel oil convert easily to gases which when burned produced light and heat; gasoline and diesel when mixed with air provided explosive power for an engine. Moreover, oil is essential as a lubricant. In his time gas storage was very difficult, and oil had the advantage that it could be stored and yet easily converted to a gas when required. In summary Frankel explained, "It is perhaps an appropriate speculation that the particular value of liquid hydrocarbons derives from their being easily gasified" (Frankel 1968, p. 13). Frankel went on to explain the disadvantages of coal or wood: "solid fuels have first to be burnt to raise steam, and only then will the steam pressure drive the engine: whereas, by using a combustible gas or material that can be gasified easily, we 'eliminate a separate energy-converting unit' and bring power to bear in the most direct form" (Frankel 1968, p. 14).

In Frankel's view, oil had superior qualities as an energy source, but its liquid state gave it unusual properties; it required specialized equipment to produce, transport, refine, and market. Typically, such equipment had extensive economies of scale and was capital intensive; it necessitated a large up-front investment, but very little labor was required once the process was in place. Because oil facilities were so specialized and capital intensive, Frankel concluded that the natural state of the industry was large-scale vertical integration: single ownership over all stages of production and distribution, concentrated in a few large companies. Less specialized economists frequently condemned his view, seeing him as an apologist for an industry notorious for its monopolizing instincts. Frankel, however, based his conclusions on the technology of his time and by the tools of economic analysis available to him then. Although the fundamentals of petroleum liquids have not changed, the number and types of competing alternatives have greatly expanded, and economic analysis has progressed and now has better insight into the problem of specialized assets.

According to Henry Ford's recollections, in 1896, Thomas Edison praised the young Ford's choice of car design: "Young man, that's the thing: you have it. Keep at it. Electric cars must keep near to power stations. The storage battery is too heavy. Steam cars won't do either for they have to have boiler and fire. Your car is self-contained—carries its own power plant—no fire, no boiler, no smoke, and no steam. You have the thing. Keep at it" (Sultzberger 2004, p. 2). In 1900, there were 2,370 cars in the three largest U.S. cities, but the bulk of them were steamers or electric cars; only 400 were gasoline-driven (Sultzberger 2004, p. 1). The internal combustion engine really took off in 1908 when Ford introduced the Model-T. It used liquid fuels with a robust engine; it could use either gasoline or ethanol (grain alcohol). Most importantly Ford built the Model-T on an assembly line at much lower cost; it was designed for a mass market. Edison struggled for over a decade to find an improved battery, but could not best the lead acid battery for cost-effectiveness and reliability. Unfortunately, the lead acid battery was too heavy to be the primary storage for an electric car, and it has taken a century for more efficient batteries to emerge. The last great impediment to the dominance of internal combustion engines was resolved in 1911 when an electric ignition system replaced the hand crank. It is more than a footnote, however, that Henry Ford did not see gasoline as the ultimate fuel source for his cars. He championed alcohol fuels into the 1930s, always wary of the natural depletion of oil.

Crude oil, gasoline, jet fuel, and diesel were not always of such central importance as they are today. In the nineteenth century, coal was far more important to industrial activity. Indeed, in the early years, refiners sought mainly to extract kerosene from crude oil. The leftovers, including gasoline, were often waste products that refiners burned or dumped, polluting rivers and streams. Kerosene was in high demand as a fuel for lamps, replacing whale oil, which was growing scarcer and more expensive every year. One of the motivations for inventing the internal combustion engine and using it to power a horseless carriage was the wide availability of gasoline and its low cost. In 1880, kerosene sold for 7¢ per gallon, while naphtha and gasoline sold for 4¢. By 1904, these fuels sold for similar prices, still very cheap, at 7¢ per gallon (about $1.76 in current dollars). In other words, oil became important largely because it was cheap and plentiful. Now, a huge infrastructure of production and distribution and an equally complex infrastructure of use—cars, trucks, petrochemical processing, and so on all depend on oil's availability. For now, consumers that own cars or trucks have little choice if they want use their vehicles to get about; they have to buy oil.

Because consumers are locked into specific types of cars, furnaces, and other equipment, the demand for oil is insensitive to price; economists say that the commodity is "price inelastic." Technically this means that when oil prices rise, consumers will cut back, but they will cut proportionally less than the relative price increase. In other words, when oil prices rise, consumers' behavior does not change very much; they just pay the higher bill, leaving them less to spend on other things. In contrast, producers can expect higher revenue and usually higher profits even though they may be producing less than before. For transportation fuels, short-run price elasticity is extremely low. When gasoline prices doubled from 2006 to 2008, drivers cut back, but only by a small amount, about

5%. The relatively small impact that higher gasoline and diesel prices have on consumption has led many observers to believe that oil is irreplaceable; something that modern society cannot do without. In fact, however, nothing could be further from the truth. Given time, there are many substitutes for conventional crude oil, on both sides of the demand and supply equation.

Price Cycles: Frankel's Nightmare

Historically, oil prices have cycled up and down in response to unexpected surges in demand or major new discoveries. The nineteenth-century oil market was especially volatile until John D. Rockefeller managed to gain control through consolidation of refining and pipeline transportation. U.S. trustbusters, however, broke up the Rockefeller monopoly in 1911. In any case, Standard Oil's combine did not extend to the producing sector, and crude oil field prices fluctuated widely. The discovery of the great East Texas oil field in 1930 caused oil to run free in the streets, it had so little value. Nonetheless, the discovery did not dampen prices for long, as demand quickly expanded.

Through much of the nineteenth and early twentieth centuries, the oil industry lurched from feast to famine. Paul Frankel (1968, p. 67) noted that the oil market was not self-adjusting: "Hectic prosperity is followed too swiftly by complete collapse." In Frankel's view, expansion of major oil companies resolved this dilemma, and a few large companies dominated the market from the 1930s to the 1970s. These companies actively engaged in every sector of the industry—exploration and production, transportation, refining, and product marketing, and they were large enough to take advantage of the industry's natural economies of scale.

Vertical integration failed to prevent the 1970s oil price shocks; however, because, as it turned out, the majority of low-cost oil resources were located in regions of the world where mineral rights are vested in the hands of governments instead of diversified private hands. This developed into a hydra-headed problem. Not only did it reduce the number of competing suppliers, increasing supply-side market power, it politicized production and pricing decisions. The turn of events manifested itself in a government-based international cartel, the Organization of Petroleum Exporting Countries (OPEC). As OPEC's power rose, the major companies' control declined. In contrast to the earlier era, OPEC's stewardship of the oil market has led to an economic disaster, spinning the oil market into a series of unpredictable price spikes and macroeconomic shocks. Frankel's nightmare returned.

The price cycles are the consequence of short-term inelastic demand and supply functions. Inflexible types of oil-dependent vehicles and equipment bind consumers to specific fuels, just as field, pipeline, and refinery capacity limits constrain producers' ability to supply. At or near full capacity, the marginal cost of producing one more unit rises dramatically. If demand growth outstrips the existing oil producing and refining infrastructure, the market requires extreme price movements in order to clear. At various times OPEC, and particularly Saudi Arabia, have attempted to stabilize the market by swinging production up and down in tandem with demand. In the mid-1980s, however, Saudi Arabia abandoned its role as "swing producer." So far, attempts by the OPEC cartel, as a whole, to manage price by production cuts among its members, or expansion in the face of rising prices, have failed. Moreover, as oil consumption soared in China, India, and other rapidly developing economies, OPEC did not expand capacity and the consequences in 2008 were a radical price increase, followed by an even more dramatic price collapse.

The Nature of the Resource

Conventional crude oil deposits lie mainly in sedimentary basins around the globe. Geoscientists generally agree that oil and gas derive from compressed and cooked organic matter that buried to significant depths over millions of years.[1] Oil migrates from source rocks and accumulates in reservoirs (commonly porous sandstone or carbonate), capped by impenetrable material (often shale or salt). Since the oil is under considerable pressure, drilling into a reservoir causes it to flow into the well bore and up to the surface. For a given type of deposit, normally the greater the pressure and the more permeable the reservoir, the greater will be the rate of flow and the lower the cost of production (fig. 2–2).

At a certain point after the reservoir is fully developed, the pressure begins to lessen or decrease and the rate of natural flow tapers off. Petroleum engineers describe this as *decline*, and the decline rate is the percent decrease in flow per year. There are many varieties of techniques to maintain pressure and moderate decline. Enhanced recovery is, however, very costly, which means that the cost of extraction rises with the age of the field. This happens for two reasons. Enhanced recovery in a mature field usually involves the drilling of new wells, which obviously adds cost. Water or gas is injected from these new wells either for pressure maintenance or for reinvigorating oil production from older, depleted wells. Added costs come from the injected materials and the energy needed to pump them into the reservoir. These costs vary widely depending on the field's age, the quality of the oil, and the complexity of the geology. Figure 2–3, describing Alaska North Slope oil, illustrates a typical profile of a major oil field's production cycle.

In Alaska's case, the discovery of the Prudhoe Bay oil field sparked a variety of additional activities. Following discovery, geologists delineated the field and estimated its oil reserves. The initial production plan put a cap of 1.5 million barrels per day on production. That production rate was expected to last until 1984, when the field would begin a natural decline, estimated at 9% per year. However, a series of extensions including new wells, a miscible gas project, and

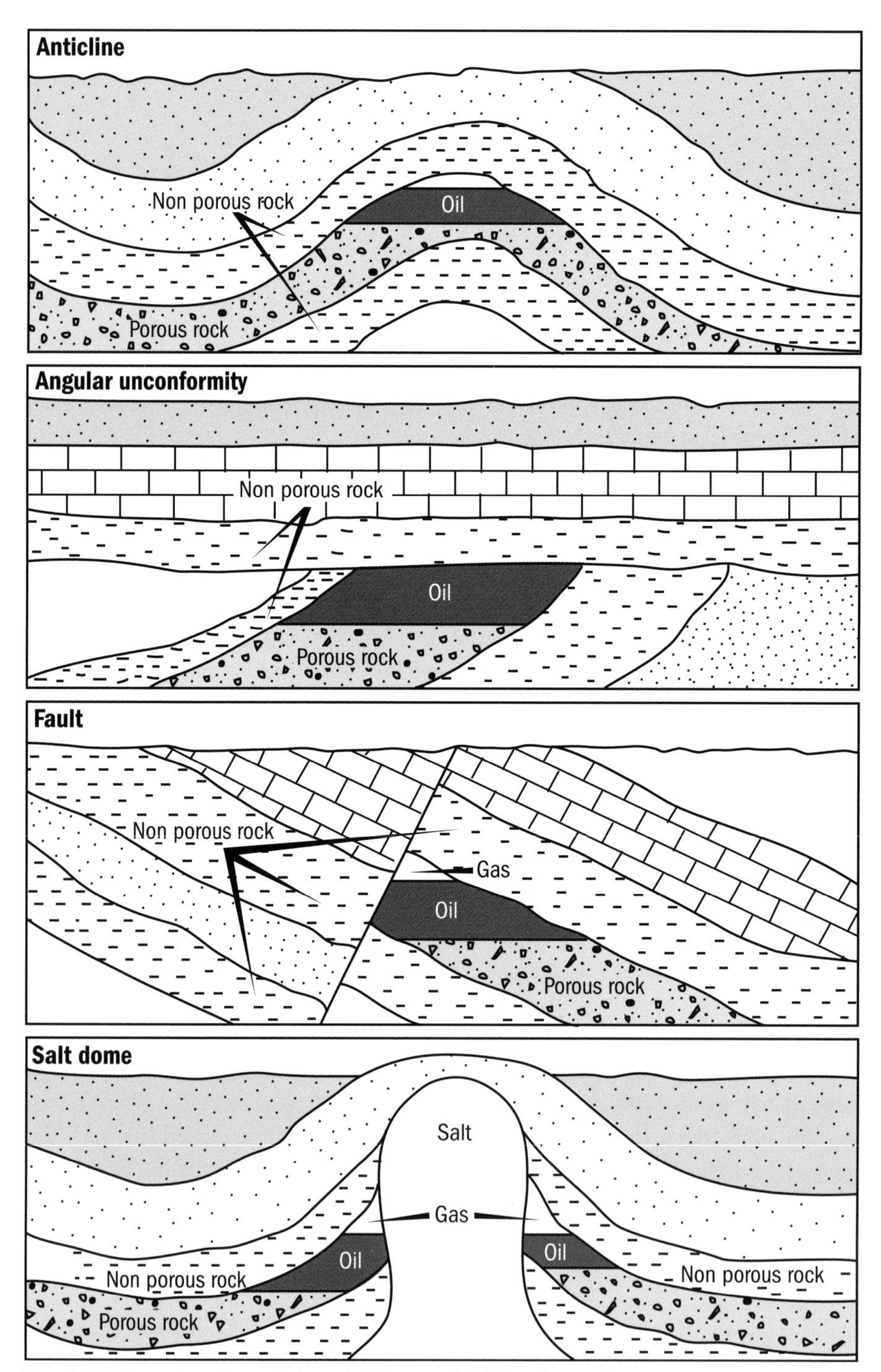

Fig. 2–2. Illustration of an oil deposit *(Source: Oil & Gas Journal)*

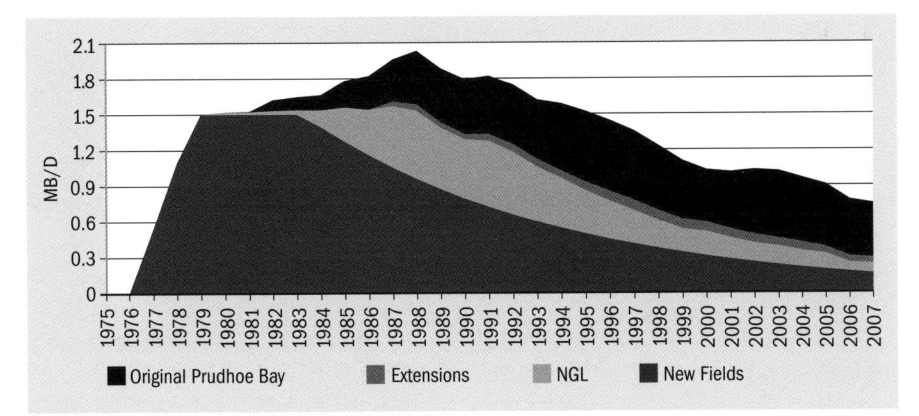

Fig. 2–3. North slope crude oil production

other activities supplemented the original plan enhancing production, as illustrated by the light gray portion of figure 2–3. In addition to the extra oil, the field also began to produce natural gas liquids (NGLs), represented by the thin band in the graph. By far the most important activity was the exploration for and development of new fields around Prudhoe Bay. The development of infrastructure greatly enhanced the incentive to explore for new fields, because the projects piggybacked easily to existing pipelines and processing plants.

Ever since the beginning of the petroleum industry, there has been a wide mix of large and small oil fields in various stages of development, producing liquids that range from "natural gasoline" to very dense tarry substances that are only liquid when heated.[2] Because there is so much diversity in fields and crude oils, it is difficult to generalize much about the cost structure and production profile, although the better the quality of the oil and the reservoir, the more likely the field will follow the expected pattern of natural decline.

The first major oil discovery in the United States was made in 1859 in Pennsylvania by drilling wells near oil seeps and other indicators. The well still operates, but only to sell oil to tourists. Most of the better quality oil fields discovered in the nineteenth century are shut down, with the wells depleted. In contrast, some of California's heavy oil (and higher cost) fields have produced oil since the 1890s. Over the years, production profiles have waxed and waned as price incentives and technologies changed. If heated, California heavy oils will flow up through conventional oil wells and into pipelines. The most common of these processes is to generate steam and pump it into the reservoir, allowing oil to flow to the surface.

Lower quality oil has a higher cost of day-to-day extraction, making supply from these fields more sensitive to prices than fields with high-quality oil. The industry expects crude oil quality to decline in the coming decades, since the best deposits were the first developed. This has implications for long-term supply price elasticity and market behavior; generally, the greater proportion of production from higher-cost fields with heavier oils, the lower will be price volatility.

In the early decades of the industry, exploration largely relied on luck. Discovery of a great field could make an explorer immediately wealthy, while "dry holes" constituted over 90% of all wells drilled and made more than a few bankrupt. Today, better knowledge about oil occurrence and geological formations, plus better technology, combine to reduce the risk of exploration; nonetheless, the unpredictable nature of discovery still drives the culture of the industry. Today, however, the focus is more on price risk than exploration risk.

Following World War II, oil exploration focused on the Middle East, where geology combined a variety of features resulting in many of the largest hydrocarbon accumulations on Earth. Of the 24 largest oil fields in the world (those found with more than 10 billion barrels,) 16 are located around the Persian Gulf. Discovery of all of these fields, including the largest, Saudi Arabia's Ghwar (or Ghawar) field occurred before 1969. Oil discoveries during this period were so vast that development had to be scaled back in order to protect the market.

The industry classifies large oil fields as "giants"—with producible reserves of over 500 million barrels; "super giants" are fields with over 5 billion barrels; and a few exceptional fields are often referred to as "elephants." For example, the largest oil field in the world, the Ghawar field in Saudi Arabia, contains at least 120 billion barrels of producible crude oil and may ultimately produce more.

Giant oil and gas fields have broader implications than simply the economies of scale they obviously provide. In 1978, Richard Nehring analyzed oil supply and reached a surprising conclusion. Despite the maturity of the industry, most oil reserves were in a few giant and super giant fields: "An estimated 819.2 of the 1,011.5 billion barrels [of reserves] are contained in known and potential giant fields. Over half the total is in the 33 super-giant fields. Roughly 10% of the total is in the estimated 20,000 fields smaller than 100 million barrels each" (Nehring 1978, p. vii). Nehring's analysis suggested new oil finds would be limited and existing resources depleted at an increasing rate. This view was frightening, and contributed to President Carter's energy policy, predicated on diminishing oil supplies, from a "wasting" resource.

Matthew Simmons extended Nehring's analysis in 2000, believing that the basic observation was still relevant. Simmons estimated that in 2000, the largest 116 fields produced 47% of the globe's oil production and it took more than 4,000 fields to produce the bulk of the remainder (Simmons 2005, p. 374). He added two other points. First, the number of giant oil field discoveries since 1970 have been significantly less than in preceding decades. Secondly, data on production rates and reserves associated with most giant fields are unreliable (Simmons 2000, p.11). In general, he is correct about both points; there have been fewer giant discoveries in recent times, except in offshore deep water. Moreover, OPEC's members have kept a tight lid on field production and reserve data. Such data that make their way into the public arena are controversial and frequently inconsistent. Although Simmons is associated with a pessimistic view of future oil supplies, he also observed: "This does not mean the world has 'run out of oil.' On the contrary, it highlights how resourceful the world has been in replacing giant fields with rapid and widespread exploration of far smaller fields" (Simmons 2000, p. 28).

A quick perusal of table 2–1 tends to validate the recent pessimism about oil supplies. There has been nothing like the string of super-giant discoveries between 1948 and 1968. Since 1980, there have been only three super-giant discoveries, and two of these have been in Kazakhstan, which was only recently opened to western oil companies. The other recent large discovery, the TUPI field in Brazil, is offshore in deepwater, has low-quality oil, and will be costly to

Table 2–1. Oil fields with estimated reserves greater than 10 billion barrels

Field	Location	Startup	Discovery	Recoverable Reserves	Remaining Reserves	Percent Depleted	Peak Production
Ghawar	Saudi Arabia	1961	1948	126,201	64,551	48.9%	5,573
Greater Burgan	Kuwait	1946	1938	48,372	20,533	57.6%	2,416
Safaniyah	Saudi Arabia	1957	1951	36,536	21,044	42.4%	1,552
Rumalia	Iraq	1954	1953	24,807	13,283	46.5%	1,534
Samotlor	Russia-Siberia	1969	1961	21,163	3,432	83.8%	3,027
Kirkuk	Iraq	1934	1927	19,853	5,098	74.3%	1,424
Cantarell	Mexico	1981	1976	17,500	6,000	65.7%	2,100
Romashkin	Russia-Urals	1945	1943	17,125	1,944	88.6%	1,081
Upper Zakum	UAE	1982	1964	16,125	16,125	0.0%	650
Shaybah	Saudi Arabia	1998	1968	14,698	13,156	10.5%	1,000
Abqaiq	Saudi Arabia	1946	1941	14,348	3,347	76.7%	1,056
Gashsaran	Iran	1940	1928	14,084	4,686	66.7%	921
Kashagan	Kazakhstan	2009	2000	13,600	13,600	0.0%	1,800
Ahwaz Asmari	Iran	1959	1958	13,597	4,290	68.4%	1,082
Lagunillas	Venezuela	1926	1926	13,140	325	97.5%	237
Manfia	Saudi Arabia	1964	1957	12,800	12,332	3.7%	1,100
Margun	Iran	1965	1964	12,173	2,101	82.7%	1,344
Khurais	Saudi Arabia	1963	1958	12,082	11,826	2.1%	1,075
Prudhoe Bay	US	1977	1968	12,015	1,266	89.5%	1,540
Zuluf	Saudi Arabia	1973	1965	11,899	8,087	32.0%	600
Northern Fields	Kuwait	1960	1955	11,692	7,156	38.8%	900
Bokan PSC	Indonesia	1964	1940	11,651	1,505	87.1%	963
Agha Jari	Iran	1939	1936	10,933	1,663	84.8%	1,023
Fyodorov	Russia-Siberia	1973	1971	10,662	2,241	79.0%	723
Total				**517,056**	**239,591**	**53.7%**	

develop. Estimated reserves are 6.5 billion barrels, only 5% of the Ghuwar field. Of the 24 fields listed in table 2–1, almost all are past peak production and on average (as of 2007,) 54% depleted.[3] Peak oil production normally occurs before the midway point of depletion. Following the peak, production declines at a steady pace, between 5% and 10% per year, unless supplemented by additional investment in new wells and other enhanced recovery activities.

The world has indeed been greatly resourceful in finding many new oil fields. Since 1973, conventional production expanded in multiple countries outside of OPEC, from a total of 24.7 million barrels per day in 1973 to 41.1 million barrels per day in 2007 according to the Energy Information Administration's *Annual Energy Review* (EIA AER). During this period, OPEC's market share slipped from 55.7% to 43.9% (EIA AER). This expansion happened despite a declining production in mature provinces, such as the U.S. With a few exceptions, however, giant oil field discoveries are mostly in deep water in the areas off East Africa, Brazil, the North Sea, and the North American Gulf Coast. These finds were possible because of substantial technology improvements since the 1970s. Figure 2–4 provides an overview of the various trends. There are a number of significant features. First, the industry enjoyed extraordinary growth in oil production and demand from 1960 through 1973 (an average of 7.8% per year). Next, there was a decline in production and demand from 1979 through 1985, all absorbed by OPEC. After the oil price collapse of 1986, OPEC production increased, returning to levels achieved in the 1970s. Overall, conventional oil supply plateaued from 2003 through 2008; the continued growth in liquid fuels demand was met more and more by unconventional sources (fig. 2–4).[4] Outside of OPEC, parts of central Asia, and deepwater in the Gulf of Mexico and off Brazil and Africa, the industry is squeezing the sponge to retrieve oil from smaller and less prolific fields, with much more intense drilling in remaining deposits.

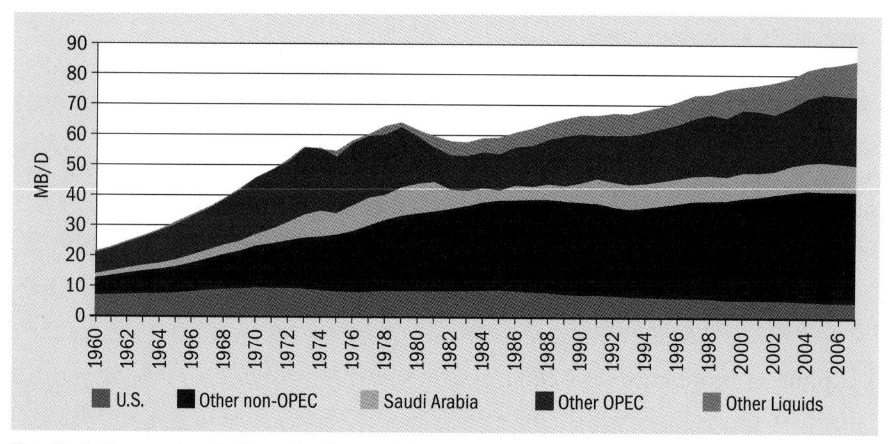

Fig. 2–4. Conventional oil production and liquid demand

The rise of OPEC was not an accident; 18 of the world's largest 24 oil fields are in member countries. Of the six remaining, three are in Russia, one is in Mexico, one is in Kazakhstan, and one is in the U.S. The sad fact is that most of the world's most prolific oil-producing regions are out-of-bounds to western oil companies. This has created a schism in global oil development. Inside OPEC, Mexico, Russia, and a few other countries, oil could be developed and produced at much lower cost than in the remainder of the globe. For a variety of reasons, including the unconscious or conscious exercise of market power, many OPEC members have chosen to hold back such development. They either develop their own resources at a slower than optimum pace or simply rely on past development, in effect, eating into their seed corn. This, in turn, has focused most western exploration to the High Arctic and deep water, where technological knowhow can be applied to resources that are harder to find and more costly to exploit.

Geographic Distribution of Oil Production and Reserves

If the surface of the globe morphed to resemble the distribution of conventional oil production or reserves it would barely be recognizable. The Middle East would dominate the globe. North America would be the next largest block, followed by Africa, East Asia, South America, and Europe. The Arctic would probably look much larger, but might shrink once explored.

One interesting feature of oil distribution is that the largest oil deposits are usually not close to population centers. Oil tends to be located in remote places and often in a harsh environment. There are a few amusing exceptions. The famous Beverley Hills High School has an oil well on its grounds (appropriately decorated with daisies and other endearments). There was something of a minor celebration when royalties from logos of Beverly Hills School 90210 outpaced oil royalties. Generally, however, oil has been hard to find, and once found it is often quite a distance from its market.

Figure 2–5 illustrates the distribution of oil production in 1973 and 2008. A remarkable feature of the distribution is how little it has changed in 34 years. The Middle East still dominates supply, but its share has dropped from 36% to 31%. North America's share has declined from 23% to 17%, reflecting the substantial decline in U.S. oil production. Oil production in Africa has increased, particularly in the Western offshore and in the Sudan, increasing Africa's share from 10% to 12%. Production in Europe and Eurasia has increased, raising the share from 17% to 22% mostly reflecting the development of the North Sea and, recently, Kazakhstan. The Asia Pacific share has increased slightly, mainly due to a tripling of production in China. South America's share of oil production has remained the same, but the numbers belie substantial change. Venezuela's oil production has declined, while Brazil's has substantially increased. It is worth noting that the comparison of 1973 and 2008 would look different if Middle East oilfields had been open to western companies. It is possible that the region could have been producing over one-half of world oil production, instead of one-third, because the lowest cost oil remains in that region.

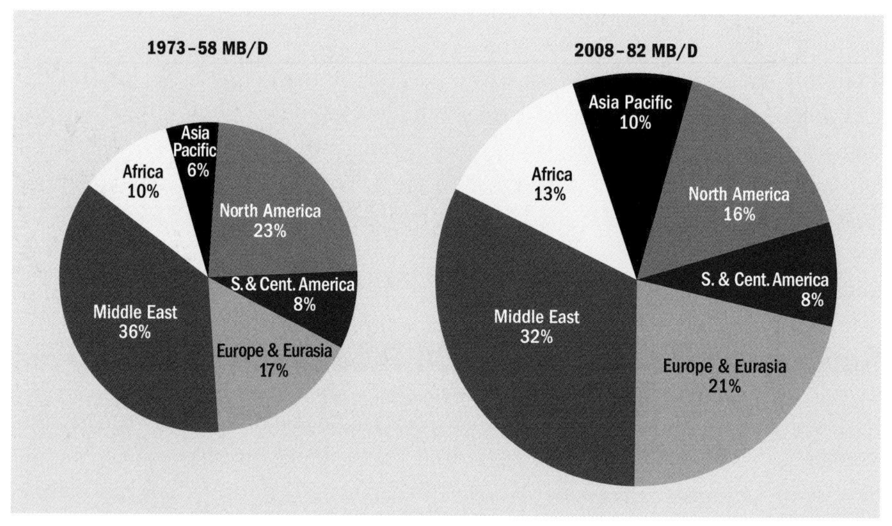

Fig. 2-5. Geographic distribution of oil production

Within the Middle East, Saudi Arabia is the largest oil producer, with an installed capacity of 9 to 11 million barrels per day; that has been more-or-less constant since 1973. *Arabian Light* crude oil, the primary output of the Ghawar oilfield, dominates Saudi production. Although Ghawar is the center point of the Saudi oil industry, the kingdom has six other super-giant oil fields that each, on discovery, contained 10 or more billion barrels of oil. The Persian Gulf has a unique geology that was perfectly suited to capture and preserve large pools of oil. Millions of years ago, it had shallow seas rich in algae, ultimately covered with a salt cap that trapped the hydrocarbon material. Nothing matches the sheer scale and scope of the Persian Gulf region as an oil province. An experienced exploration geologist was once asked if another Saudi Arabia would ever be discovered. He replied, "yes, but it will be in Saudi Arabia."

There are multiple sources of oil production data by country, and both government agencies and the trades press follow the figures closely. There are a few measurement issues. A barrel of oil in a pipeline, marine tanker, or storage tank is a barrel of oil, In contrast, data on oil reserves are much more speculative and in some instances purposely distorted. Much of the apparent oil reserve increases logged in the 1980s and 1990s are from OPEC countries. Colin Campbell (1997) pointed out that OPEC's members had an incentive to inflate reserve estimates, because reserve levels influenced production quotas. In any case, close examination of reported oil reserves on a country-by-country basis reveals more than a few anomalies. In some countries, the numbers remain the same from year to year, sometimes they are rounded, and in a few cases, there are significant increases without reports of new oil field discoveries.

These problems arise because there are no external standards for data reported by OPEC and many other countries. Publicly traded energy corporations report production and reserves to standards laid down by the U.S. Securities and Exchange Commission. These standards arose because in the early twentieth century it was too easy to mislead stock buyers by overstating reserves. Even within these standards, however, there is a great deal of flexibility. Deutsche Bank (DB) provided an example of a North Sea gas field where multiple companies involved in a joint project gave reserves estimates that varied by more than a factor of 2 (DB 2008, p. 78). Without public scrutiny of oil production and reserve data or without a common understanding of reporting standards, it is prudent to be skeptical of any oil supply forecast. Unfortunately, that skepticism translates to market uncertainty, which may be partially responsible for the extreme price volatility observed in 2008.

Perhaps the largest single problem in reserve estimates is the easy way in which they may be misinterpreted. Reserves are by definition only discovered resources producible under current technological and economic conditions. Reserves are not an estimate of the total amount of oil likely to be recovered. From an economic perspective, it makes little sense for a company to develop reserves beyond a 10-to-12-year horizon. Instead, they set about to replace the reserves on their books. In other words, they seek to discover and develop at least enough reserves to replace the oil and gas produced during the year. The best example is North American natural gas, which has had a reserve life of approximately ten years for over three decades. Scarcity enthusiasts frequently cite low reserve figures as evidence of imminent decline. By themselves, however, reserve figures provide no such corroboration, as newly established reserves continually replace the produced gas.

In retrospect, it is apparent that the oil shocks of the 1970s were not a precursor of imminent supply decline. The next two decades saw the development of oil supplies throughout the world and an outright increase in "published reserves" over and above production during the time-period. Moreover, much of this supply came about in an era of falling oil prices and from new fields all over the world. It was OPEC's members, however, who booked the largest reserve increases, and the cartel's retreat into isolation raises serious questions about the data's reliability. Among other things, suspicion about the validity of reserve estimates stirred a lively debate about when and why oil production will peak.

The Peak Oil Controversy

Hubbert's Peak

In the 1950s, the noted geologist M. King Hubbert predicted the peak and decline of U.S. oil production using a technique that is now favored by a growing number of scarcity enthusiasts.[5] Briefly, Hubbert used a simple statistical technique to predict peak oil production in the U.S. (which turned out to be reasonably accurate). Hubbert began by noting that oil fields went through periods of production increase, plateau, and finally decline. He postulated that the first discoveries were small fields. Then, improved exploration techniques led to the discovery of larger fields. Finally, as exploration matured and discoveries petered out, only small fields remained to discover and develop. The sequence of all these discoveries would create a bell-shaped production profile with a defined peak. If the first part of the bell could be trended, then the second part, the decline, could be predicted by simply reversing the trend. This is known as "Hubbert's Peak."

There are several key points about the peak oil theorem. First, the methodology assumes a fixed stock of recoverable oil reserves, so if the pace of oil consumption raises, production time contracts. Figure 2–6 illustrates the trend; the solid line is a facsimile of Hubbert's original projection for the U.S. This projection is the most popular because it came close to correctly identifying 1971 as the date when U.S. oil production would turn down. The light gray dots are actual production of conventional crude oil through 1996 and a Hubbert-style projection by Colin Campbell made in 1997. Campbell has since updated his estimates, but the 1997 projection is important because it rekindled a debate that began in the 1970s. The crude oil surpluses of the 1980s and 1990s had eliminated most concerns about supplies. Campbell assumed a slightly greater amount of recoverable reserves than Hubbert, but production is still constrained. Since more oil was produced than the curve projected in the 1970s and 1980s, the rate of decline after 1997 has to be greater in order to make up the difference. In true Malthusian style, the bigger the feast today, the greater the famine to follow.

Peak oil advocates emphasize that they are analyzing only conventional crude oil supplies. Certain heavy oils, NGLs, and so on. are not considered. When unconventional oil resources are included in the analysis, the picture dramatically changes. For example, the resource base of oil shale in the U.S. is staggeringly large, but prohibitively costly to produce. If these resources could be included it could expand recoverable reserves of liquid fuels by an order of magnitude and extend U.S. resources through this century. In any case, since oil production has turned down, the U.S. has supplemented its liquid fuels supply by adding ethanol, biodiesel, and so on. Figure 2–6 illustrates the difference in

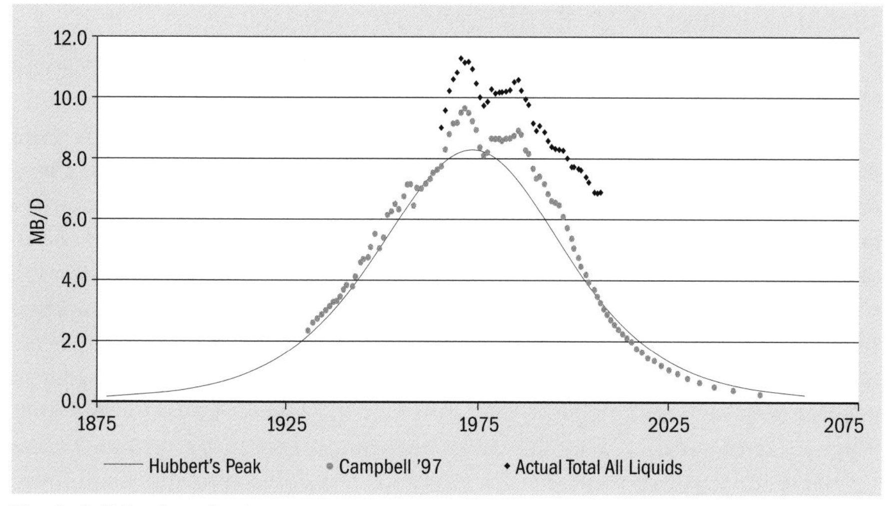

Fig. 2–6. U.S. oil production

assumptions by the black dots. Campbell's 1997 methodology predicted that U.S. oil production for 2007 should have been 3.3 million barrels per day, one-third of its peak. Actual production of all oils in 2007 was about 6.9 million barrels per day.[6]

Given reduced opportunities to find conventional oil, the industry has shifted emphasis. Recently the Energy Information Administration of the U.S. Department of Energy (EIA) changed its description of basic fuel types from oil or petroleum products to "liquids" (EIA 2007, p. 6). Crude oil is still the major source of liquid fuels, but NGLs, synthetic oil (from tar sands,) ethanol, and gas to liquids supply sources are growing in importance both in the U.S. and in other regions. Likewise, the major oil companies no longer publish oil reserve estimates in their annual reports; instead, they report barrels of oil equivalent (BOE), which includes unconventional resources and natural gas. The shift to unconventional petroleum resources, however, has a significant impact on the industry's cost structure, which will be higher but proportionally less capital intensive.

Although Hubbert's projection was wrong about the extent of supplemental oil supplies in the U.S., it was right about an inevitable decline of conventional oil production. If the U.S., with its market-based system, cutting-edge technology, and abundant resources could not arrest the decline, what does it portend for world oil production? The peak oil methodology may be hopelessly naïve, but it does encompass an indisputable fact: oil resources in the earth's crust are finite and cannot be relied upon forever.

Global estimates

During the period of scarcity in the 1970s, analysts occasionally cited Hubbert to underscore the pessimistic view of future oil and gas supplies for the entire globe.[7] However, Hubbert did not actually reveal the specifics of his formulae until 1982 (Campbell 1997, p. 93). During the 1980s and 1990s, a large amount of spare oil production capacity in OPEC and the former Soviet Union emerged, which kept prices suppressed, leading the industry and the public to expect increasing production and modest prices.[8] Thus, Colin Campbell showed some courage in 1997 by resurrecting the idea of another oil crisis amidst a swamp of surplus. Among other things, he utilized something like Hubbert's methodology to predict the global production of crude oil. Campbell predicted a "price leap" beginning in 2000 and a plateau of production until about 2009, when "conventional" crude oil production would begin an inexorable decline. To the amusement of some, he also included a chapter entitled: "Economists never get it right." Although Campbell was right about the price leap, he was wrong about the timing of the production constraint; ironically, right about the economics, but apparently, wrong, or at least premature, about the geology.

Campbell is not alone in ringing the oil supply bell. Since 1997, a growing number of geologists and laypersons have become fascinated with the peak oil concept. Part of the fascination lies with Hubbert's compelling image. For environmentalists or anyone suspicious of the industrial age, the bell curve is a striking symbol of impending doom. Professor Deffeyes of Princeton University further enhanced the respectability of the analysis in his 2001 book *Hubbert's Peak*. Deffeyes, like Campbell, estimated the global production peak between 2003 and 2009 depending on total recoverable reserves that he and others have estimated to be between 1.8 and 2.1 trillion barrels. Deffeyes went much further in a later magazine interview, suggesting that peak production would be on Thanksgiving Day in November of 2005.

Perhaps the most vocal of modern-day alarmists has been Matthew Simmons. In his book *Twilight in the Desert*, Simmons undertook a detailed review of the Saudi Arabian oil industry. According to Simmons, almost everyone expected the Saudis to not only continue to produce oil at about 10 million barrels per day, but to increase production by 50% to 100% over the course of the next decade. After his study of the Saudi's super-giant fields, Simmons concluded that this would not be possible. He added to the controversy by claiming that production from the Ghawar oil field had likely peaked at around 5 million barrels per day and was in permanent decline. Saudi officials have vehemently denied Simmons's analysis. Nonetheless, despite increased investment activity and high prices, Saudi Arabia's oil and NGL production has yet to exceed the 2005 average of 11.1 million barrels per day (BP 2009).

One of Simmons's major points is indisputable. Since the rise of OPEC and the withdrawal of the major international oil companies from key producing areas, oil field information and investment plans are opaque, which makes planning and investment in remaining regions more risky and difficult.

In contrast to the peak oil view, energy use changes slowly, but in time it does change. Since the oil crisis of the 1970s, energy demand has continued to grow, while oil has lost market share. Total demand in 1973 was the equivalent of 115 million barrels of oil per day; in 2008, energy demand had nearly doubled to 226 million barrels per day. The share of oil in energy consumption slipped from 48 to 35%, while natural gas increased from 19 to 24% (fig. 2–7). Over the coming decades this trend is likely to continue, because the natural gas resource base appears to be significantly greater than that of oil.

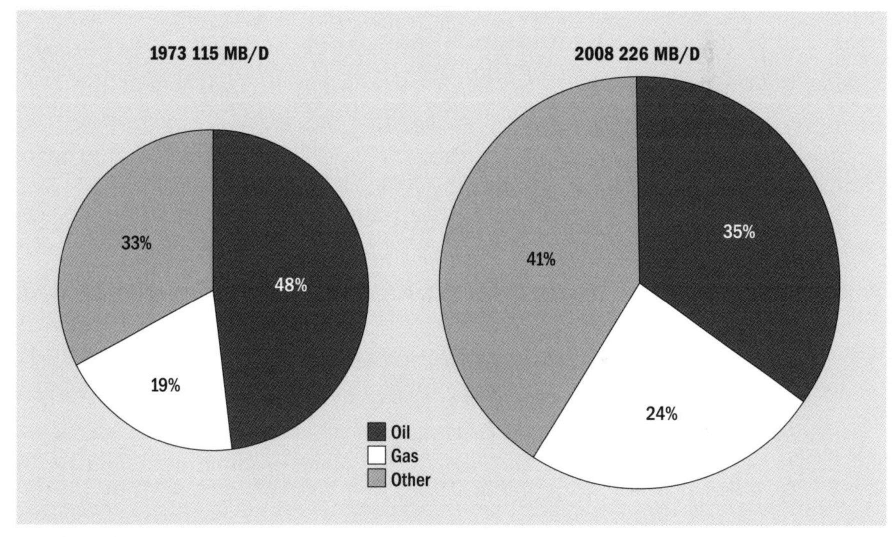

Fig. 2–7. Changing market share of energy use

As Kenneth Boulding observed decades ago, "Infinite growth on finite resources is not possible." In the broadest sense, all resources are finite, if limited to the earth. However, economic growth does not necessarily depend on increasing volumes of physical goods, nor does increased energy use necessarily depend on fossil fuels. Chapter 8 provides a series of visions concerning future energy systems.

Alternative views

Michael Lynch and a host of economists have criticized both the notion of an imminent peak in oil production and Hubbert's methodology, which quantifies it. Lynch's criticism of the peak oil forecast methodology is convincing.

He notes that initially Hubbert adopted the bell curve as a descriptive device, but that it has morphed to a prescriptive or "explanatory" tool (Lynch 2003, p.1). In the view of peak oil advocates, the downturn of oil production is a fact of nature, inevitable and unalterable by economic incentives. In contrast, most forecasts, such as those prepared by the EIA, see a substantial increase in liquid fuels over the coming decades, although it will be a mix of conventional and synthetic oils.

So far, conventional wisdom, as cast in the public announcement of oil companies, government forecasts, and private forecasts, dismisses the notion of an imminent peak in oil production. There are two main reasons given: first, the experience of the last two decades has demonstrated vast opportunities in oil resource development all over the globe. In the forecasts of the 1970s and 1980s, analysts overlooked much of this new supply. It was easy enough to track the large projects in the North Sea, Alaska, etc., but small discoveries and incremental adjustments are easy to miss and can add up to a large quantity. Thus, many analysts were shocked to find that from 1973 to 2007 oil production in developing countries had increased over threefold, from 4.0 to 13.9 million barrels per day (BP 2009).[9] Most forecasters had thought this segment of the market would do well to stay even; they were dead wrong.

Second, as already explained, peak oil theorems are greatly constrained by considering only conventional oil supplies. They also assume virtually no technological improvements. From a historical point of view, many new types of oil discoveries were due to trial and error. Following the discovery of a new type of deposit, geologic theory adjusts to explain it. A classic example is the discovery of the East Texas oil field, which Professor Deffeyes described. "East Texas, the largest oil field in the lower 48 states, is located where the Woodbine Sandstone was tilted up, partially eroded, and buried by younger sediments. When East Texas was discovered by accident in 1930, this type of trap was not on the agenda of oil geologists" (Deffeyes 2001, p. 45).

There is some evidence to suggest that the East Texas discovery was something more than an accident. The explorer, "Dad" Joiner, had oversold subscriptions (like the characters in the movie, *The Producers*) and he was probably drilling in the *least* likely place to find oil. More recently, Thomas Gold tried to verify his theory of the abiotic origins of oil by drilling into a basalt formation in Sweden. The drilling discovered traces of oil that critics claim could have been from the lubricants of the drilling bit. In any case, scientists now believe that Saturn's moon, Titan, has lakes of liquid methane, which supports aspects of Gold's theory, at least with respect to natural gas. (In a reversal of perspective, oxygen would be fuel on Titan). There is no doubt that some form of hydrocarbons exist in vast quantity throughout the universe. To assume that today's geologists know everything there is to know about future oil discoveries is simply naïve.

The Staircase of Rising Cost

Cost diversity

If drillers knew nothing about the size and location of oilfields they would most likely find the larger fields first. Indeed, Professor Deffeyes reports that the Russians became frustrated with the analyses of their geologists and simply laid out a horizontal grid for drilling. Apparently, they had just as much success. The intuition of this observation is obvious; the greater the horizontal extent of a field the higher the probability it will be found before any of the smaller fields. This is also the logic behind the first inflection point on the rising portion of Hubbert's Peak. As the industry learns how and where to drill, it has greater and greater success; eventually, however, big finds taper off and, as they do, production increases can be expected to decline, and eventually the supply of oil will dwindle.

Large oilfields are by far the cheapest to exploit. In addition to their horizontal breadth the deposit usually has a greater vertical depth. This means that the field will likely require fewer wells per barrel, the overall production rate will be higher, and the oil flow will last longer. These are all features that reduce cost. Major oil companies (the big IOCs) normally focus on finding giant oilfields, where they can exploit the economies of scale offered by their size and expertise. Once a field has aged past its most prolific years or if it is too small, the majors sell the properties to independents, who specialize in smaller scale operations. There are thousands of unproduced oil deposits left standing, because they are too small, or the geology too complicated.

The huge difference in the size and quality of known oilfields means that there are significant cost differences in finding, developing, and producing conventional crude oil, and costs rise with uncertainty. The risk and cost of finding and developing the Prudhoe Bay oilfield in Alaska illustrate this issue. Before an oil company can drill a prospect, it must purchase the property or obtain a lease for the mineral rights. In the case of Prudhoe Bay, the land was owned by the State of Alaska (the State was granted large chunks of federal land when the territory became a state). Alaska chose a portion of the coastal plain in the High Arctic between the Brooks Mountains and the Beaufort Sea. The choice was not without controversy, because the only value of the land was its potential for minerals or oil development. Alaskans referred choice to as "Marshall's icebox" or "Marshall's folly," named after the geologist that made the decision. As luck would have it, the land Alaska chose contained the largest oilfield ever found in the U.S. along with a whole set of satellite fields and enormous natural gas reserves.

State and the federal governments have similar policies when leasing mineral rights. In the U.S., governments usually grant leases through a competitive auction—an up-front or bonus payment that gives the winner the right to explore the leased area. Following discovery and development, the leaser usually pays a royalty, in-kind or in-value; the royalty rate is typically 12.5% to 15%. There are often constraints to the drilling rights. For example, if the company does not drill in the first few years of the lease, rights revert to the leasee. Alaska's key auction of North Slope acreage was held in July 1967 and garnered a bonus of $6.1 million for 754 thousand acres. Two years later ARCO (then Atlantic) discovered the Prudhoe Bay oil field. As it turned out, however, BP had acquired acreage in the same lease sale that ended up amounting to about one-half of the field's oil. The two companies agreed to jointly develop and operate the field, with BP taking the eastern half and ARCO taking the western half.

Following the Prudhoe Bay discovery in September 1969, Alaska held an additional lease sale for 461 thousand acres for land surrounding that offered in the July 1967 sale. This time the bidding went ballistic and the State made a total of $900 million (Alaska DNR 2007, pp. 1-4). In a final twist of irony, after Prudhoe Bay was developed, bidders learned the first set of leases contained 97% of the Prudhoe Bay field. The second auction's "winners" wound up with only about 3%.

In characteristic fashion, the Prudhoe Bay discovery spawned exploration and new development activity. In the ten-year period following the 1969 lease sale, companies drilled 100 exploratory wells on state lands and 19 discovered oil or gas. The largest new find was the Kuparuk River oil field, which has produced over 2 billion barrels of oil out of an expected total of 2.8 billion barrels. In addition, three other fields (or areas), Milne Point, Endicott, and Point McIntyre, are expected to exceed 500 million barrels of production and qualify as giant fields (Alaska DNR, 2000 and 2007).

Some of the exploration effort has been a spectacular failure. Since 1979 the federal government has held a series of lease sales for the outer continental shelf (OCS) of the Beaufort Sea, to the north of Prudhoe Bay. There was tremendous excitement about the 1982 lease sale, and in total the industry spent $2.1 billion for the right to drill. There was particular interest in one part of the acreage, a specific prospect nicknamed "mukluk." Three companies, headed by BP, spent nearly $1 billion in drilling costs and lease fees in order to drill one well. That project may be the most expensive dry hole in history. Among other things, the partners had to build an island in 60 feet of water to do the drilling (Bradner 2005). According to the head BP geologist, this prospect was the only structure large enough on the North Slope to contain another oil field the size of Prudhoe Bay. Unfortunately, the drilling revealed that the cap rock was faulty and the oil had migrated elsewhere. Nevertheless, federal and state offshore leasing has continued on the North Slope since then, and four fields have been discovered; three, Kuvlum, Hammerhead, and Sand Piper, proved uneconomic and were not developed. The fourth prospect, known as Liberty, is currently under development but is unlikely to qualify as a giant field.

Although there is great cost diversity from one conventional oilfield to another, almost all are significantly lower in cost than the unconventional alternatives. The problem is that conventional oilfields are limited in number and geologists have found most of them. Peak oil advocates may not be able to peg the date (or dates) of the peak with any accuracy, but everyone agrees that the resource is finite and that at sometime in the future conventional oil production will begin to decline.

The remarkable feature of conventional oilfields is the ease of recovery of minerals deep in the earth's crust. This is because the fields are under substantial heat and pressure. When a drill bit broaches a deposit, pressure pushes the oil to the surface. Today, gushers do not spill on the ground as they did in the early twentieth century, but they are under considerable pressure. This means that the production or lifting costs in the early life of a field are minor. Typically, conventional oil is costly to find and develop but once developed it can be produced at a relatively low cost, at least up to the field's capacity. Put another way, conventional oil is highly capital intensive. In contrast, unconventional hydrocarbon resources are not only higher in total cost; they have greater day-to-day production costs because extraction requires mining or the active injection of steam or chemicals (fig. 2–8).

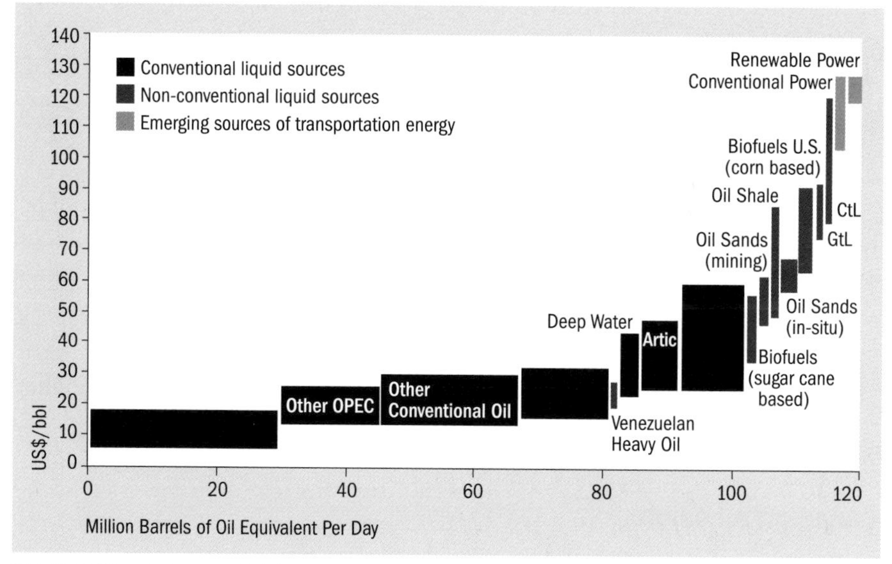

Fig. 2–8. The staircase of rising cost

Ed Morse of Barclays has prepared a chart of the various conventional and unconventional options for liquid fuels production. The categories of fuels, their cost, and expected production levels change from time to time, but the chart provides a clear illustration of the increasing staircase of cost. As discussed earlier, liquid fuels have a distinct advantage for the transportation sector. Unlike oil, however, the natural state of most unconventional fuels is either solid or gas. They require conversion to a liquid if they are going to be useful. This requires an extensive capital investment and a significant amount of energy just for the conversion alone. Unlike conventional crude oil, the net energy produced by these options is lower, and virtually all raise the level of greenhouse gas emissions. The conversion makes economic sense if the final product is sufficiently superior—the alchemist trick of turning lead into gold. This is not always the case, however, particularly if part of that industry segment is subsidized.

The staircase of cost is multidimensional, encompassing an investment of both time and money. If the technology is still unproven, it must undergo a period of research and development (R&D), prototype development, and finally a scaling-up sufficient to make it cost effective. Companies experimented with many of these technologies in the 1970s and 1980s, with some significant failures. Exxon and Tosco attempted to produce oil from oil shale with the Colony Project in Colorado. They abandoned the investment in 1982, after a $400 million dead-weight loss (DeMott 1982). In contrast, the U.S. Department of Energy (DOE) finished the Great Plains coal gasification project with an investment cost of approximately $2 billion. The plant was a technical success, but commercial failure; it has produced pipeline quality synthetic gas almost continuously since 1984. However, the original sponsors backed out of the plant and turned it over to the DOE, because of the high cost. Eventually DOE sold the plant to a utility, without recovering the bulk of the investment. Both of these projects cratered primarily because the price of oil began a steady decline after 1980.

When both the Colony and Great Plains projects got underway, global warming was not yet an issue. Thus, emissions from the conversion process concerned air quality, not the double impact on greenhouse gases of converting a low-grade hydrocarbon and then burning the reprocessed fuel. Recently the Great Plains' owner modified the project to add a pipeline that delivers CO_2 to oil fields in Saskatchewan, becoming the first energy facility to sequester such emissions (DOE OFF 2006). A discussion of emissions from conversion technology continues at the end of the next section.

Canadian oil sands

The province of Alberta has a huge set of unconventional oil deposits, referred to as oil (or tar) sands. There are three primary locations: Cold Lake, the Peace River, and the Athabasca River. Support infrastructure is located at nearby Fort McMurray. Much of this resource lies near the surface and can be strip-mined, which is how development began. The oil sands themselves are a mix of clay, sand, water, and a tar-like substance, referred to as bitumen. The Oil Sands Discovery

Centre estimates that the total bitumen in place is from 1.7 to 2.5 trillion barrels. In 2007 Canadian oil sands reserves, based on firm development plans in place, were increased to 8.9 billion barrels. The potential, given present technology and expected prices is, however, much higher. Canada's National Energy Board (NEB) now posts total "established" reserves at 173 billion barrels. According to NEB Board Member Roland George, present production is about 1.3 MMB/D, but in 2008, he expected it to ramp up to between 2.6 and 4.9 MMB/D by 2030, depending on oil prices, technology, and environmental constraints (George 2008). The most recent NEB (2007) "reference case" put total production at 2.8 billion barrels in 2015. Since the oil price decline, however, many of the oil sands projects have been suspended or scaled back.

The industry has known about the oil sands resource for decades, but the high cost of development prevented any substantial exploitation. The first venture, known as the Great Canadian Oil Sands (now Suncor) began production in 1967. Low oil prices, however, kept the project at a low level of development. Syncrude followed the Suncor project in 1974. It was a joint venture of eight Canadian and U.S. oil companies, including Petro Canada, Imperial Oil (Exxon), and ConocoPhillips. Suncor and Syncrude are the two biggest producers and expected to add the most capacity over the next few decades. After oil prices peaked in 1979, oil sands continued to develop, but at a slow pace, and the emphasis shifted from strip mining to "in-situ" production.

The in-situ production method is similar to that used to produce California heavy crude oils. The most common approach is "steam assisted gravity drainage" (SAGD). As the name implies, the process relies on heating the bitumen with steam in order for it to flow after drilling two horizontal wells. One well pumps steam down and the other collects bitumen and pumps it to the surface. This process has a lower capital cost than surface mining, but requires considerable energy to generate the steam. Much of the recovered bitumen is upgraded on site, and this too is an energy-intensive process. The final product in the initial Syncrude process is a low-sulfur diesel that moves easily through conventional pipelines.

As oil prices rose from 1999 to 2008, interest in the oil sands grew and so did expected costs. According to the NEB, costs increased about three-fold to mid-2008, reflecting a stronger Canadian dollar and the global cost escalation experienced by all sectors of the oil and gas industry. Since then, costs have fallen (table 2–2).

Table 2–2. Oil Sands Economics 2008, Canada NEB

	CAPEX $C/bbl	**Threshold WTI US$ bbl**
Mining, extraction, and upgrading	$90k–$100k	$55–$65
In-situ	$28k–$33k	$50–$60

The high cost of oil sands did not seem to put a damper on the industry's enthusiasm, in part because royalty rates (around 1%) have been very low. A primary motivation for major oil companies is to replace declining reserves that they carry on their books, which is important in setting expected future profits and, therefore, stock prices. Although the oil sands development is costly, there is little or no exploration risk. Once committed, the project's reserves can be booked to the company's account. However, at the daily per-barrel cost most recently cited by the NEB in 2008, it will cost at least $100 billion to add production capacity of 1.5 MMB/D by 2015, plus high operating costs once facilities are in place.

Aside from price risk, the oil sands projects also face growing environmental opposition. Fort McMurray does not have the "NIMBY" ("not in my backyard") problems associated with projects in major urban areas, but there is concern about the projects' impact on greenhouse gas emissions. The CO_2 produced from generating steam and upgrading the bitumen may require additional investments to sequester the emissions that could reach as high as $C 16 billion (*Petroleum News* 2008). Retrofitting comes with an even higher price tag.

The emissions issue underscores the major problem with tar sands and virtually all conversion technologies: they are energy-intensive with corresponding side effects. At present, the Oil Sands industry plans to use natural gas as the primary fuel to generate steam. However, the in-situ process will require from 2 to 2.6 million Btu of natural gas to produce and upgrade one barrel of synthetic crude oil. In 2007, NYMEX natural gas prompt month prices averaged $7.09 per million Btu. Futures oil heating oil prices, on the other hand, averaged $14.78 per million Btu. Thus, the industry expected to use a cheap fuel to manufacture a more valuable one. The two-to-one value relationship may not hold in the future; previously the price of natural gas fluctuated widely with respect to oil, but in the long-term averaged the thermal equivalent of heavy fuel oil prices.

Gas-to-liquids technology

Gas-to-liquids (GTL) technology is a refinement of the Fischer-Tropsch process developed in the 1920s and used by the Germans in World War II to turn coal into oil. This technology has been adapted to natural gas to produce a low sulfur diesel. The conversion costs of gas to oil are about $20 per barrel, but the overall total cost is highly sensitive to the price of the feedstock, particularly since the GTL process consumes approximately 20% of the gas in conversion. A number of companies have successfully developed the GTL technology; SASOL and Royal Dutch Shell are the most advanced. During the period of South Africa's isolation, the country's energy policy aimed at self-sufficiency, and its national oil company, SASOL, converted coal to oil. By the mid 1990s SASOL had improved its process to the point that it could produce oil from coal for under $30 per barrel (about $40 in today's money). These figures are somewhat misleading, because South Africa has an abundance of low-cost coal, which would not be available in other regions. After one major setback, Shell developed a successful gas to oil facility in Malaysia. The company is now constructing a huge GTL plant in Qatar; the facility will produce 140,000 barrels per day of synthetic oil and 120,000 barrels per day of liquefied petroleum gas and condensate (Shell Web site). Other companies are experimenting with the technology, but the major drawback is the dependence on a hydrocarbon feedstock and the large amount of energy required in the process.

Oil shale

Another staggeringly large resource is oil shale. The largest concentration of this resource is in the Green River basin of the Rocky Mountains, which may contain the equivalent of up to 1.8 trillion barrels of oil (Grunewald 2006). Like oil sands, oil shale does not flow at room temperature; heat is required to separate the oil from the encasing sediment and other impurities. The hydrocarbon in oil shale is karogen, an early stage of conventional crude oil. (In contrast, the hydrocarbon in oil sands is bitumen, which is a late stage of crude oil, where the lighter ends have dissipated.) The two primary methods of extraction for oil shale are surface mining followed by retorting and in-situ, where the resource is heated in the ground. Both methods, however, have significant environmental side effects. Although Exxon's oil shale project failed, oil shale converts to oil in other parts of the world, primarily Estonia. Oil shale burns like coal, but it is a messy process. Shell operates a small in-situ test facility on private land in Colorado, producing high quality oil and gas from the site.

Renewable fuels

The development of renewable fuels has been full of controversy. Biofuels including corn ethanol are the most talked-about renewable alternatives to oil, but there have been serious accusations that the total energy required to produce corn ethanol exceeds its net energy content. If those claims are correct, then replacing oil with corn ethanol may actually increase rather than decrease dependence on foreign imports and exaggerate the global warming problem. Ethanol production, mainly from corn in the U.S., receives the largest subsidy of any energy source—approximately $4.7 billion in 2006, 26.5% of the total spent on the product (Texas Comptroller 2007). Perhaps most importantly, the rise in oil prices accompanied an almost one-to-one increase in grain prices in 2008. Since grains are a vital part of the human food chain the impact of persistent higher prices on humanity's poorest could have tragic consequences.

There is no doubt that the production of ethanol by conventional distillation requires a great deal of energy. Fuels are used to plant and harvest the grains, the distillation process requires heat and, of course, transportation and blending is required to produce a marketable product. Concern over the "net" energy of corn ethanol reached a peak in 2005 when two scientists from Cornell University and the University of California Berkeley published a study indicating that: "To produce a liter of ethanol requires 29% more fossil energy than is produced as ethanol..." (Pimentel and Patzek 2005, p. 66). Other studies before and after dispute the authors' conclusion. Although there is debate over a variety of calculations, the primary issue is how to account for all the energy inputs, given that many of the activities to produce ethanol support multiple objectives and make products other than ethanol. For example, waste products from distillation are used as animal feed.

To some extent the emphasis on net energy is misplaced. To pick an extreme example, producing a diamond requires a great deal of energy, but produces none in return. Yet, a diamond still has great value for entirely different reasons. To extend the analogy, ethanol plays an important role in improving fuel quality; when mixed with gasoline it enhances octane, improving cylinder compression and reducing engine knock. Furthermore, because it contains a substantial amount of oxygen, it reduces carbon monoxide emissions, which is particularly important in winter. Alternatives to ethanol have serious side effects. For many decades the industry added tetra-ethyl lead to gasoline in much smaller proportions (and much lower cost) to achieve the same octane results. Tetra-ethyl lead is, however, highly toxic and was likely responsible for increasing the general level of lead in the environment. Furthermore, the lead interfered with catalytic converters, installed to improve car emissions and reduce smog. Consequently the U.S. chose to phase it out as an additive in 1973. Refiners added methyl tertiary butyl ether (MTBE) in the 1980s—serving two purposes, as an octane enhancer and oxygenate. The Environmental Protection Agency (EPA,) however, prohibited MTBE when it began showing up in water supplies. So far, ethanol has proved to be the least harmful option for meeting fuel quality requirements consistent with protecting the environment.

U.S. ethanol production currently receives two forms of financial support, a direct subsidy to corn farmers, and a tax credit for ethanol content in gasoline. It is likely that these subsidies were necessary to develop the industry and ethanol has proven to be beneficial. Nonetheless, expanding the use of ethanol beyond about 10% of total fuel demand has diminishing benefits to the environment and does not necessarily reduce dependence on foreign oil, or at least not very much. Indeed, to the extent that the subsidies drive up corn prices and impact all grain prices it could prove detrimental to global economic development and in particular to the impoverished. It may be cheaper for the U.S. to import additional sugar-based ethanol from countries like Brazil, rather than expand production of corn from prime farmland at the cost of other grain production. Distillers claim a side benefit in that growing corn sequesters CO_2. Recently a group of scientists challenged this claim, arguing that excessive tilling in the U.S. provokes deforestation around the globe and net impact is an increase in greenhouse gases (Searchinger, et al 2008).

The other primary option for renewable fuels is biodiesel. Biodiesel comes from recycled cooking oils or agricultural products such as soybeans, corn, palm trees, and the cotton plant. A catalyst is added to the vegetable oil to produce a fuel, which can substitute for diesel in most cases. The use of recycled cooking oil has obvious efficiency benefits as do agricultural products that do not compete with the human food chain. Constrained supply is an obvious problem; the amount of oil that can be recycled is very small relative to oil consumption.

There has been a humorous side to the rise in oil and grain prices in 2008. Aside from rising bread and rice prices, which were not funny, Germany had a "beer" crisis and Italy a "pasta" crisis. There is no doubt that grain prices rose in tandem with oil prices through 2007 and 2008, and the increased production of ethanol and plant oils, are partially to blame. The exact causal relationship is yet unclear, but the global economic boom was the root cause of higher prices for both.

Producing cellulosic biofuels, rather than relying on grains for fuel would, of course, minimize the impact on the price of rice, wheat, corn, and other edible plants. In this process, waste material—corn stalks, woodchips, and other vegetable matter is fermented or otherwise converted to a biofuel. There is active R&D aimed at finding enzymes and other additives that would make the process more efficient. Many believe that these technologies hold promise, but as of 2009, there are no commercial plants producing biofuels from cellulous material.

The U.S.'s Energy Independence and Security Act of 2007 (EISA 2007) sets a highly ambitious goal for ethanol and biodiesel production. The act requires fuel producers to use at least 36 billion gallons (or 2.35 million barrels per day) of biofuel by 2022. According to the Energy Information Agency this would represent 17% of U.S. liquid fuels consumption in 2022 (EIA 2008, p. 82). It would also represent a staggering production increase of 510% from the 5.9 billion gallons of biofuels used in 2006. Twenty-one billion gallons of the 36 billion gallons of biofuel slated for use in 2022 are supposed to be from "advanced" biofuels (not grains or edible plants), but this seems impractical given the present technology.

At least so far, renewable fuel costs do not set or determine oil prices; they reflect them. This is mainly because energy costs are a large proportion of renewable fuel costs. As oil prices rise they tend to pull up fertilizer prices and all other fuel prices, usually required for production and conversion. In addition, both oil prices and agricultural product prices tend to follow the business cycle, which means they generally rise and fall together. Renewable energy prices and costs will always be towards the top of the staircase of cost, but the absolute value of cost associated with their production will vary.

The Perils of Forecasting

There are several rules most forecasters observe: Round off all the projections to at least four significant digits to show a sense of humor. For a short-term forecast, connect the last two observations and extend the trend with a ruler. For a long-term forecast, use a yardstick. Be careful to predicate the forecast on a set of equally unpredictable assumptions so that errors require no explanation. For example, sages that predicted oil prices to reach $150 per barrel by the end of 2008 assumed robust economic growth through the year. It did not seem to cross anyone's mind that the two events might be mutually exclusive. No matter how sophisticated the modeling effort, past trends drive forecasts, because the only way to sense the future is to examine the past. Generally, this procedure works well—the future often mimics the past. Trend analysis, however, misses significant turning points due to unexpected technological, social, or economic events. Thus, had the East Texas oil field not been discovered, oil prices would have been more likely to have followed past trends, but the discovery changed everything.

The forecasts by the EIA for the U.S. government, the International Energy Agency, and a number of private forecasters are in sharp contrast to the gloomy paranoia of peak oil enthusiasts. Indeed, until the 2007–2008 oil price run-up most forecasters saw conventional oil supplies peaking after 2030 at about 100 million barrels per day.

Most peak oil advocates believe that government and industry forecasts of conventional oil supply are faulty. They point out that the forecasts depend on a healthy increase in OPEC production, from 36 to 49 million barrels per day. This, in turn, depends on a very substantial increase in Saudi Arabia's oil production (an increase from around 10 to 15 million barrels per day). The controversy swirls around the issue of oil field decline and the reliability of data that supports the competing analyses.

One way to form an opinion about future oil supplies is to compare current thinking to previous forecasts. In the 2004 International Energy Outlook, the EIA forecast world oil (liquid) production of 121 million barrels per day in 2025 based on a price of slightly over $30 per barrel (in 2006 USD). In 2008, the EIA forecast total liquid production of 107 million barrels per day in 2025, at prices that averaged around $80 per barrel (in 2006 USD). This is a significant difference, but more important it is simply inconsistent with sound economic theory. If oil prices more than double there should be more, not less, daily supply. Of course, in the four intervening years the community of oil analysts had become increasingly pessimistic about future oil supplies. With lower prices, the opinion may reverse again. Put another way, current events sway the forecasting community more than they should and their projections of long-term supply reflect current perceptions rather than sound facts. Despite the sophisticated modeling, the results are little better than using rulers and yardsticks.

According to armchair historians, that there are only three great lessons from history: First, "who the gods wish to destroy, they first make mad with power." Second, "the wheel of the gods grinds slowly, but it grinds exceedingly fine." And, of course, the third observation, revised by famous litigator Clarence Darrow: "History repeats itself. That's one of the things wrong with history."

It is instructive to review similar oil market forecasts made in the past by different individuals and institutions with a very different agenda. In the late 1970s, it was fashionable to forecast a gap between rapidly growing demand and laggard supply. The purpose of the gap was to encourage policy makers to do something about the energy crisis. In this instance, the forecasters were purposely depicting an infeasible future. Table 2–3 summarizes these forecasts and compares the results to what actually happened.

Table 2-3. Comparison of 1985 Oil Production and Consumption

Organization	Demand	Supply		
		Non-OPEC	OPEC	Gap
CIA	70.5	21.5	44.2	5.3
EIA/DOE	68.1	23.8	39.8	4.5
CRS	67.8	25.0	42.8	–
Exxon	64.0	24.0	40.0	–
PIRA	61.0	24.4	36.6	–
Pet. Economist	61.0	25.5	35.5	–
Actual	**41.4**	**25.0**	**16.4**	**–**

From forecasts published in 1977 and 1978 (in million of barrels per day, net of FSU, China)

Surprisingly, the forecasts of future non-OPEC supply were reasonably accurate, although they stretched less than a decade ahead. The forecasts of oil demand, however, were wildly high and it was lower oil demand that closed the gap, not increased supply. It is now possible to put this error in perspective. Oil demand had grown very rapidly in the 1960s due to low prices caused by the massive reserve discoveries in the Middle East. Consumers viewed oil as the "swing fuel." As long as prices were moderate, oil supported most incremental economic activities. Given time and high oil prices, however, consumers substituted alternative fuels, particularly in the boiler fuel and the home heating market. Since the 1970s, oil has retreated to the transportation sector, which now accounts for three-quarters of demand.

Uncertainty about future oil supplies and the failure of energy forecasts to predict the key changes of the last few decades boils down to a simple conclusion:

forecasters know far less about the future of oil supply and demand than they are willing to admit. Supply data for the OPEC members are not reliable and will not likely improve. More importantly, there may be new technologies or discoveries on the horizon that will completely alter perceptions about the future, as similar advances have in the past. Chapter 8 explores some of these technologies and the implications for oil prices.

Notes

1. A competing concept of petroleum origin, known as the "abiogenic theory," postulates that hydrocarbons form in the Earth's mantle from primordial methane, at considerable depth, and then migrate into the upper crust. This theory enjoyed wide currency among Soviet geoscientists, and remains very much alive in Russia today. It was championed in the U.S. during the 1980s and 90s by Thomas Gold of Cornell University, but is not generally accepted by Western scientists. A major implication of the abiogenic theory is that petroleum resources are vastly larger than commonly estimated and that great volumes exist at depths well below the current reach of the drill bit.
2. The industry categorizes crude oil by its density and the extent of contaminants. Density is measured by American Petroleum Institute Gravity (APIG). Crude oils in the range of 34° to 44° APIG are described as "light" and are considered the most suitable for refining because they make a high proportion of gasoline, diesel, and jet fuel, the most valuable products. In contrast, "heavy" crude oils, 9° to 14°, make mostly asphalt and heavy fuel oil unless sophisticated refining procedures are employed. The most common contaminant in crude oil is sulfur and the greater the proportion the greater the refining cost and/or the lower the value of products produced. Crude oil prices vary significantly depending on location, APIG, and sulfur content.
3. Some care must be exercised in interpreting various sources of oilfield data. Since data on specific fields has been increasingly difficult to obtain. Moreover, the principal sources of such estimates are often inconsistent.
4. A precise accounting of the amount of "conventional" and "unconventional" sources of oil supplies is very difficult. This is because natural gas liquids are increasingly used to supplement crude oil production and many refineries have a volumetric gain because they convert crude oil to less dense products.
5. This author was lucky enough to hear a presentation by Mr. Hubbert in the 1970s.
6. There are slight differences in data depending on the source. "Actual" oil production used in Figure 3–1 is from BP statistics with an estimate for 2007 based on the percentage difference from DB.
7. Hubbert followed his U.S. forecast with a world view; it was not, however, as accurate in part because it did not anticipate the 1980s drop in demand growth.
8. OPEC's attempt to maintain high oil prices in the early 1980s resulted in a reduction in oil consumption and at times nearly one-half of the cartel's production capacity was closed down. Lower prices after 1985 stimulated demand, but the collapse of the Soviet economy in 1990s resulted in another wave of oil surplus that lasted until the end of the decade.
9. This category is defined as countries not in OPEC, not in the OECD, and not part of the former Soviet Union.

OIL PRICING

Price Theories of Depleting Resources

Professor Walter Mead once jested that any well-coached parrot could explain oil prices—they are set by demand and supply. Somehow, that explanation does not cut it with government policy makers, industry executives, and commodity traders. They would like a deeper analysis, and short of finding a soothsayer with a crystal ball, they are usually disappointed. Day-to-day oil prices are notoriously unpredictable, flung about by political events, changing economic growth, OPEC meetings, revised resource assessments, and consumer reactions, all of which are difficult to forecast in advance. As explained in chapter 2, both short-run oil supply and demand functions are highly inelastic, setting the market up for extreme price volatility. The long run, however, is quite different; oil prices are more predictable than realized by the popular imagination. On the supply side, the important determinants are resource availability, risk level, interest rates, and extraction technology. On the demand side, there is economic growth, technologies of use, and, most importantly, the cost of alternatives.

There are two competing theories to explain oil price determination, and both flow from similar market fundamentals. The modern understanding of oil price formation derives from Harold Hotelling, who worked out the mathematics of the relationship in 1931. His fundamental concept was that depleting resources—most commonly oil and other minerals—have a fixed inventory, with their production and use spread out over time. In its simplest form, the trade-off between using the resource today and saving it for the future depends on current and future demand, the total inventory of the resource, market interest rates, and the expected cost of a replacement, referred to as the *backstop technology*.

Figure 3–1 is a greatly simplified condensation of Hotelling's theory, with two different results. The first state of the world, denoted by P_1 and Q_1, begins by assuming a backstop technology with a cost in the future higher than current oil prices. (In this diagram, oil has no production costs.) The current price, P_1, is determined by discounting the backstop technology cost to the present using market interest rates. In the second state of the world, P_2 and Q_2, there are two changes. First, the expected cost of the backstop technology is higher. Normally, this would increase the present value of oil, thus causing current prices to rise. However, the illustration also assumes that interest rates have increased. Due to higher interest, the present value of backstop cost drops. If the impact of higher interest is greater than the increase in future technology costs, current oil prices will drop. Lower oil prices, however, mean greater consumption. This reduces the amount of time for inventory depletion, moving the development of the backstop technology closer.

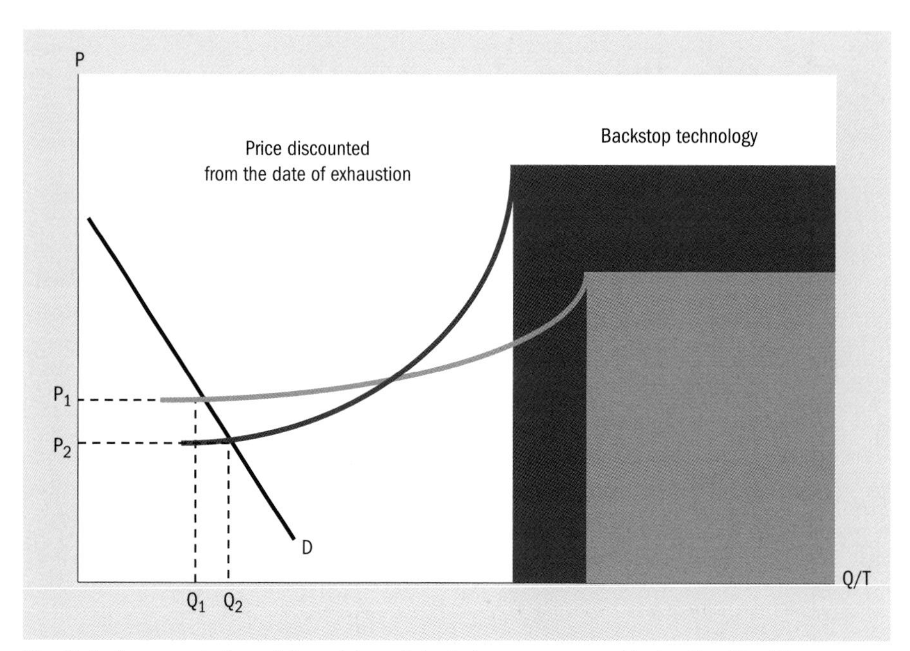

Fig. 3–1. Representation of the pricing of depleting resources with the Hotelling Theory

Hotelling himself worked out a number of nuances regarding the model—for example adding in extraction and development costs. Since his time, a number of economists have expanded the theory to accommodate a variety features. The key idea, however, is that the value of oil (or any depleting resource) is determined by a complex tradeoff between present and future use. Put another way, from the point of view of a resource owner, the marginal cost of selling oil today is not just the day-to-day cost of extraction or lifting costs, it is the discounted revenue foregone of selling the resource sometime in the future.

In some respects, the Hotelling model is comparable to the worldview of peak oil advocates; both approaches assume fixed oil supplies. However, under the Hotelling paradigm, demand in relation to price and cost in various time-periods, with an eye toward the future cost of replacement, determines production, rather than an iron rule of geology. Peak oil advocates assume that humanity, like the grasshoppers of fables, use up the oil as quickly as discovered and then society falls apart in disarray. Most economists take the Hotelling view; high prices due to shortages will simply provoke alternative energy resources. As folklore tells it: necessity it the mother of invention.

The problem with the Hotelling construct is not the theory; just about everyone agrees he got that right. The problem is the uncertainty about the key variables; no one knows for sure how large the inventory of oil really is, what will be the cost of a backstop technology or the cost of future oil extraction, or even what the appropriate discount rate ought to be. Nonetheless, the theory has been influential, particularly among OPEC members who knowingly or unknowingly summarize Hotelling by stating that oil in the ground is worth more than the investment alternatives that might arise from selling it today.

In contrast to the Hotelling approach, Professor Morris Adelman believes that treating oil as an "exhaustible resource," with the implication of rising prices over time, is misleading. He comments: "No mineral, including oil, will ever be exhausted. If, and when, the cost of finding and extraction goes above the price consumers are willing to pay, the industry will begin to disappear. How much oil is still in the ground when extraction stops, and how much was there before extraction began, are unknown and unknowable. The amount extracted from first to last depends on cost and price," (Adelman 1996, p. 11). To illustrate the point, Professor Adelman draws a graph almost identical to Hubbert's Peak oil. He explains that rising production, followed by a peak, and then decline, describes sales of 33⅓ rpm phonograph records, IBM mainframes, and a variety of products (Adelman 1996, p. 13). The crucial element of Adelman's theory it the recognition that as conventional oil fields are depleted, the cost of finding and extracting new finds will rise. The inventory of oil is not fixed; it just becomes more costly to produce until no one is willing to buy.

Figure 3–2 illustrates the production cycle as Adelman sees it. The chart describes the production of electricity using oil-fired generators in the United States. From 1966 to 1979, the amount of oil-fired generation increased dramatically. Then, with higher prices (and policy changes), oil-fired generation fell just as dramatically from 1979 through 1986. It is important to note that the amount of oil used in power generation did not decline due to a drop in electricity demand. Rather, it was a shift to other generation sources—coal, nuclear, and natural gas. In other words, scarcity and high prices provoke alternatives and substitution. On this point both Hotelling and Adelman agree.

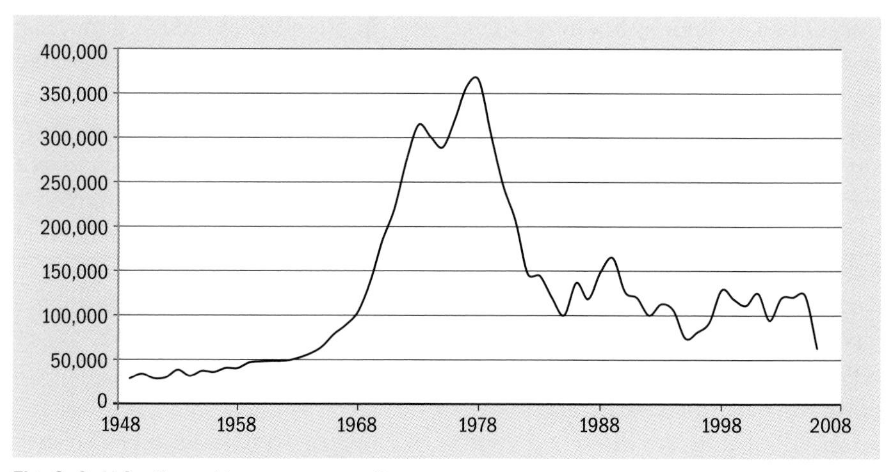

Fig. 3–2. U.S. oil used in power generation

Generally, the Adelman view of the industry tends to dominate at the top of the price cycle, such as the summer of 2008, when excess demand stretched production capacity to its limit. In contrast, when slack capacity emerges and there is downward price pressure, the longer-term Hotelling view comes into play. In the downside of the cycle, prices seldom fall all the way down to lifting costs (which they would if oil had no alternative value in future). At some point, prices get low enough that individual producers decide to close down and save the oil for another day. They may also coordinate such activities to avoid what Rockefeller characterized as "ruinous competition." OPEC's apologists have often claimed that the cartel's coordinated actions are to the long-term benefit of consumers, since they are preserving vital resources for future generations.

The Hotelling and Adelman theories merge by simply noting that the extraction of oil begins with low cost fields. Many such fields may have an economic rent based on the present value of the backstop technology or the higher cost of smaller oil fields. As oil becomes harder to find and the cost of extraction rises, eventually marginal development costs match the backstop technology; then prices stabilize and oil use phases out. This way of looking at the world, however, depends on diversified ownership of the low cost fields and open access to resource development. If the control of the most prolific oil fields and undeveloped resources rests in the hands of a few, the orderly path of exploitation may become decidedly rocky. That has indeed been the case, where most of the prolific oil fields are in OPEC member countries, and further investment has been limited and uneven.

Returning to Professor Mead's parrot, the bird has wised up. Now it is able to explain a bit more about what determines the key functions of oil demand and supply. When there is surplus capacity, an increase in interest rates should lower oil prices, because financial investments will have a higher return than leaving oil in the ground. On the other hand, higher interest rates raise the cost of expanding capacity setting up a possible price rise when demand overtakes supply. An expected increase in the cost of replacing oil sometime in the future will raise oil prices in today's market. A drop in expected economic growth will put downward pressure on oil prices prolonging the lifespan of the oil resource and increasing the odds that there will be temporary periods of excess capacity. In turn, better than expected economic activity will raise prices, particularly if there has been insufficient infrastructure investment.

It is easy to overlook a key point. Since the paradigm is dealing with a long-term tradeoff between future replacement costs and present use along with highly inelastic short-run demand and supply functions, subtle changes in expectations or new circumstances can produce dramatic shifts in observed prices. In other words, there are so many factors that determine oil prices it is often hard, if not impossible, to sort out all of the influences. In such an ephemeral and ever-shifting environment, institutional arrangements matter, and over the decades, the industry has developed both private and public arrangements focused on stabilizing oil prices when spare capacity emerges. OPEC's production quotas are one example and there is a discussion later in the section on OPEC's role in setting prices.

Marker or Benchmark Prices

When a commodity is heterogeneous (for example, if it varies in quality), pricing normally proceeds by establishing a "marker" or "benchmark" to compare and grade all other varieties. One of the most interesting examples is the diamond market. Diamond D is a one-karat stone of traditional round cut, perfectly clear, no hint of color, and no flaws. Experts grade individual diamonds and then price them relative to Diamond D. The industry tracks the price of the Diamond D stone as it fluctuates, and the price of every other diamond moves up and down in tandem with the benchmark.

As noted earlier, crude oils range in quality from natural gasoline to tar and production sources vary from Siberia to the tip of South America. Thus, the various grades, at various locations, can take on significantly different relative values. The oil market has always had certain benchmark crude oils that were widely watched by the industry to provide an index for general price movements. After OPEC assumed pricing control, it set Arabian Light f.o.b. at the Arab Gulf as the cartel's marker crude oil, shifting the pricing basis from the Gulf of Mexico to the Middle East. Arabian Light was the marker price until 1986, when the OPEC pricing scheme collapsed.

In 1983, NYMEX chose West Texas Intermediate (WTI,) for delivery at Cushing, Oklahoma, as its benchmark for the crude oil futures contract. Over time, NYMEX crude oil trading increased dramatically and WTI became scarcer.

Consequently, NYMEX broadened the contract specification to "light sweet" crude oil with delivery at Cushing. The generic description allowed traders greater latitude in fulfilling the contract in the event that physical delivery was required. As a practical matter, WTI and NYMEX's light sweet crude oil are interchangeable in benchmarking U.S. crude oil prices, since WTI is the largest component of the NYMEX group. At times, oil prices at Cushing have deviated somewhat from broader world trade, due to the land-locked location. Such deviations have made little difference to NYMEX crude oil futures trading, which remains the largest oil market in the world.

The North Sea oil market developed in the late 1970s and was one of the first physical markets to use "over-the-counter" (OTC) forward trading. Brent crude oil, which is actually a blend of light low-sulfur oil from 20 different fields, was the principal crude oil. The oil lands at the Sullom Voe oil terminal at the Shetland Islands in Scotland. In 1988, the International Petroleum Exchange (IPE), now the Intercontinental Exchange, chose Brent as the basis for its futures contract, and the new market formalized the OTC forward trading. Brent crude oil prices are widely reported, and it remains the most important benchmark for European refining and for the relative pricing of many foreign crude oils.

It is important, however, to stress early on that the heavily traded futures contracts at NYMEX and ICE are what the industry terms *financials* as contrasted to *physicals*. That is, the primary purpose of the financial contracts is to manage price risk; it is not for the purchase or sale of physical lots of the commodity.[1] Futures, options, and swaps are intended to be settled for cash, but to be successful they must be closely linked to the physical market. That is, traders selling oil in a futures contract must be willing to deliver it if they have to. For this reason, prices determined on futures exchanges correspond directly to prices that govern the physical flow of oil, and when analysts discuss crude oil prices, they do not usually make a distinction. A more detailed discussion of energy derivative markets follows in chapter 4.

OPEC's Role in Setting Prices

OPEC's pricing methodology from 1973 through 1985 depended on Saudi Arabia to act as swing producer. Initially the cartel tried to manage pricing and production for each of its members. It was reported that OPEC developed a complex computer model (on an early-days mainframe computer) that took 24 hours to work out the various quality and transportation adjustments; cynics noted that by the time the job was over the market had changed. By luck or design, the cartel hit on a methodology that solved the problem, at least for a while. The idea was to fix the price of Arabian Light, with the kingdom's production moving up and down to accommodate shifting demand. As quid pro quo, the other cartel members would agree to abide by strict production limits, but set their own official prices to reflect appropriate market-based levels. As long as no one cheated, the scheme worked well (fig. 3–3).

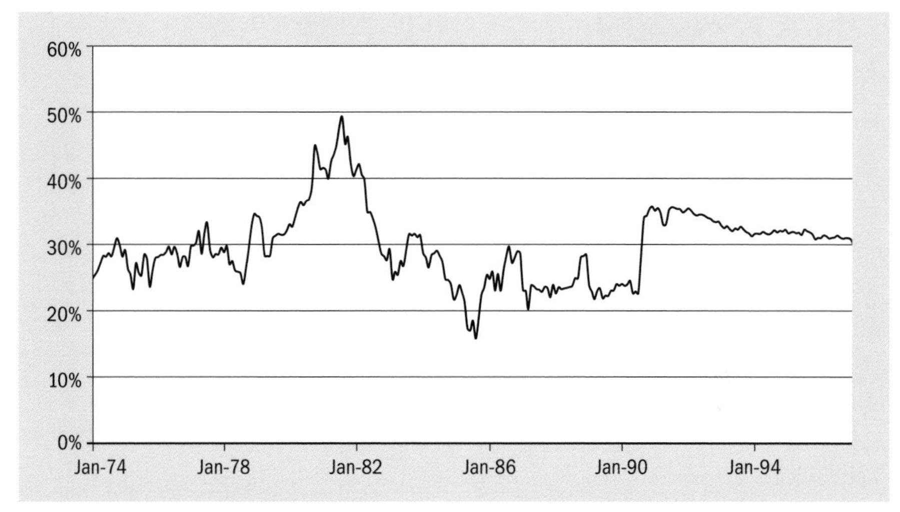

Fig. 3–3. Saudi Arabia's share of OPEC

OPECs system fell apart after the Iranian Revolution. Anxiety over oil supplies in 1979 and 1980 quadrupled spot prices, tempting the cartel to set official prices that were unsustainable. High prices, a painful recession, and new suppliers eroded OPEC's market and drove many of the cartel's members to ignore production limits. Consequently, Saudi Arabia's oil production plunged at an accelerating rate. In mid-1985, the Saudis abandoned their role as swing producer and tied the price of their crude oil to a market basket of petroleum product prices, referred to as "netback" pricing. The basic methodology was simple. Crude oil prices were determined by adding a normal refinery margin to a market basket of petroleum prices after the oil was refined. Refiners liked the idea because it guaranteed a profit and Saudi sales volumes recovered. The market stabilized for a brief period, but once the heating oil season of 1985–1986 was past, prices collapsed. It is worth noting that if the downward trend in consumption had continued, Saudi Arabia would have had to buy crude oil (rather than sell it) in order to maintain the price.

Although the formal system of OPEC price setting ended in the summer of 1985, the significance of Arabian Light as the marker price had been eroding since 1981. In a series of maneuvers, the cartel reduced the official sales price (OSP) of Arabian Light from its peak of $34 per barrel and adjusted relative prices, attempting to redress the imbalance in production between its members.

In the endgame, the adjustments failed to rectify the inequities, and Saudi Arabia abandoned the system. The NYMEX marker replaced the Saudi market and the new focus was on exchange trading.

The idea of a cartel is to fix prices higher than a purely competitive market would allow. The dilemma is that the group's interest as a whole usually diverges from the self-interest of its members. While the group may do better with higher prices and lower volumes, an individual is better off to expand volume beyond the allocated quota, keeping the action secret. In plain terms, cartels usually fall apart because the participants cheat on their obligations through price chiseling or off-the-books production. To be successful, a cartel usually has a dominant supplier and/or some sort of enforcement mechanism. With a few exceptions, most commercial cartels have not lasted more than a few years. OPEC is an exception; it celebrates its 50th anniversary in 2010. It is also exceptional in many other ways; it is an organization of sovereign states with little in common, except oil resources. In fact, Iraq has been in deadly conflict with two other members, Kuwait and Iran, at various times during the cartel's history.

Cartel watchers describe OPEC members as either "hawks" or "doves." Hawks, such as Iran and Venezuela, have limited reserves and a strong need for revenue to support a large population. They have an obvious motive for high prices and less concern about the long-term viability of their market. Doves, such as Saudi Arabia and Kuwait, are at the opposite end of the spectrum; they have long-lived reserves and less immediate need for revenue. In the cartel, the hawks argue for policies directed at high prices, and the doves are more concerned about stability and the long-term market for oil. Since this characterization surfaced, circumstances have changed for many OPEC members. Matthew Simmons, for example, argues that Saudi Arabia's population has increased dramatically over the last two decades and that the kingdom requires much higher revenue than before. Although there is some merit in this argument, Saudi Arabia needs less than $50 per barrel to cover its annual expenditures—this is much lower than the budget requirements of many other OPEC members.

From a historical perspective, OPEC has managed to garner higher than competitive prices primarily because a core group of countries has been willing to cut production when necessary and turn a blind eye to other members' cheating. Saudi Arabia's dominant position is the primary reason for OPEC's success. In 1985, for example, Saudi Arabia produced only 78% of its production quota, while every other country (except Iran due to the aftermaths of the revolution) was well above.

Since 1986, OPEC has managed its market by setting production allocations or quotas for its members and, on occasion, setting target prices or price ranges. Figure 3–4 illustrates the cartel's history of quotas, as compared to production. The figures never track exactly because there are differences in definition between crude oil production, NGLs, and other nuances. Some members avoided production quotas, Iraq being the most common exception. Data in figure 3–4 exclude production figures for OPEC members that do not have quotas. Although production has virtually always exceeded the quota total, cartel members have generally followed the scheme.

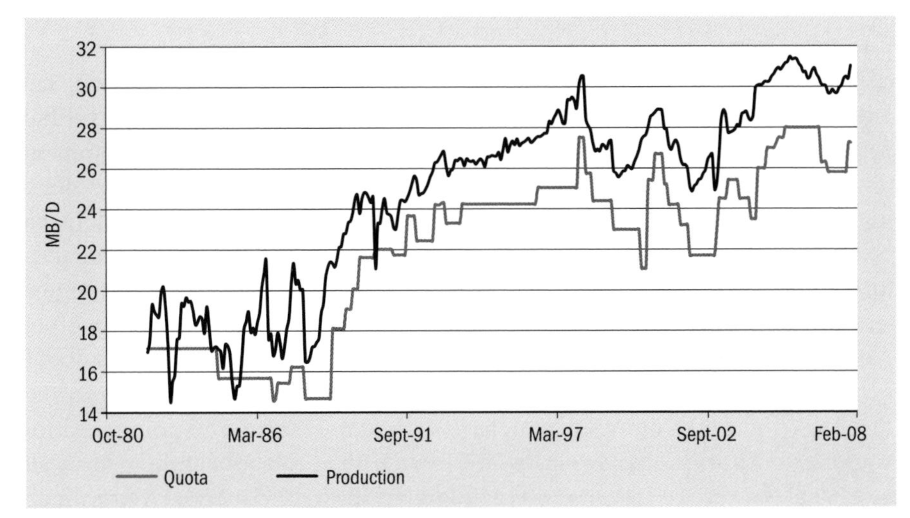

Fig. 3–4. Comparison of OPEC production and quotas

In the mid 1980s, when the market collapsed, there was considerable production volatility and consequent deviation from the quota. This stands out in figure 3–4. During this period, the Saudi's shift to netback pricing had an unplanned consequence—it demonstrated to other OPEC members the importance of the swing producer. As mentioned, for a cartel to be successful it must have an enforcement mechanism, and it turns out that OPEC's devise rests with its largest supplier, Saudi Arabia. The 1986 price collapse bottomed out on March 31 with the NYMEX prompt price reaching $10.42 per barrel.[2] From that time forward, Saudi Arabia had a credible threat. If other cartel members ignored their quota, the kingdom could flood the market and bring oil prices down to single digits. From time to time, the Saudis threaten to ramp up production, and at all times it has been an implied lever on other producers' behavior.

Although OPEC has been successful in implementing quotas that reflect demand for its oil, it is unclear whether it leads or follows the market. Econometric studies suggest that there is a negative relationship between the stringency of quotas and oil prices (Kaufman 2003). However, the greater the impact of quotas on prices in the short term, the greater the incentive to cheat, meaning that the price impact wears off quickly. There are also circumstances where the announcement of quotas is completely ineffective. For example, On December 17, 2008, OPEC decided to lower quotas by 4.2 million barrels per day. The action was not enough; prices fell. In any case, it is clear that the crudeness of OPEC's decision process has not been successful in stabilizing oil prices. In 2008, alone, prices ranged from $145 to $36 per barrel.

OPEC sets an expansive agenda for itself. "OPEC is a permanent, intergovernmental organization, established in Baghdad, Iraq, 10–14 September 1960... Its objective is to coordinate and unify petroleum policies among Member Countries, in order to secure a steady income to the producing countries; an efficient, economic and regular supply of petroleum to consuming nations; and a fair return on capital to those investing in the petroleum industry." (OPEC 2008) Over the years, the cartel has sought to meet much broader objectives than just the maximization of revenue. From their point of view, they are seeking to stabilize the market, just as U.S. programs in the 1950s and 1960s established a price floor for domestic producers. In addition, OPEC has maintained programs to help developing nations, aid the shift to alternative energy, and encourage technology transfer to further resource development.

OPEC's stated agenda and its inconsistent actions raise key questions about its real intent. Is it a cartel seeking to maximize profits or is it attempting to stabilize oil prices, promoting genuine cooperation between oil producers and consumers? There is a commonality between OPEC's members and American oil producers in the early years of the industry. Both have suffered from highly volatile prices and both have sought to develop institutions that put a floor price on oil by forcing production cuts in weak markets. Unfortunately, OPEC has been far less effective in establishing price ceilings, in large measure because there has not been enough investment in swing production capacity. As residual supplier to the world, they need the flexibility to shift supply in response to seasonal changes and economic events. OPEC is now a much smaller percentage of total global energy, and oil demand has proven to be more volatile than expected. If OPEC expects to prevent extreme price cycles, it must maintain a high level of spare capacity and have the discipline to prevent price spikes. The cartel must also be realistic about the level of prices that are sustainable in the long term. Chapter 8 analyzes many of the alternatives to oil and price levels at which they become cost effective.

Posted Prices

OPEC did not invent their system of production management and official prices out of whole cloth; they took the idea of "pro-rationing" from the Texas Railroad Commission and pricing techniques from the North American system of "posted" prices. Traditionally, North American refiners list prices for the various crude oil grades they seek to purchase. The posting bulletins were mailed to interested parties. (They are now online on the companies' Web pages.) Posted prices are for oil delivered in the field or at specific pipeline junctions or delivery points. The prices are specific to crude oil produced from a particular field or a blend of crude oils where the blend is stable and commonly understood. Field examples are Midway Sunset or Kern River oils in California. Blend examples are West Texas Intermediate or Alaska North Slope (ANS). Posted prices are "offer" prices. Prices in royalty or purchase contracts typically average postings from several companies or, in recent time, state a differential from the NYMEX futures market.

It is a myth that "posted" refers to tacking price offers to posts in the field. In an interview given by Mr. Platt in the 1950s (the founder of *Platt's Oilgram*,) he explained that the posting terminology derived from early oil market trading in the nineteenth century. At that time, there was a crude oil exchange in Pennsylvania. Posting referred to the chalkboard prices offered by refiners and asked by producers. In 1935, *Time* magazine retrospectively described the action: "Oil exchanges were exciting markets during the last quarter of the nineteenth century. The Oil City, Pennsylvania, Exchange boasted a hectic day in 1884 when 29,006,000 barrels of crude were sold. But trading was precarious when crude prices, notoriously unstable, often jumped from $2.75 a barrel one year to $20 the next." (It seems almost like the current market). The article went on to explain the plan to exchange oil in a New York futures market. The market failed, however, and the posted price procedure dominated North American markets until the rise of NYMEX.

Although posted prices are less significant pricing tools today, they are still important for determining relative crude oil prices, both for particular fields and quality differences. Typically, the postings list specific crude oil streams and a price adjustment per unit of API gravity. That is, the heavier or denser the oil, the greater the price-discount; the exact amount of the discount varies from day to day and is listed in the bulletin. If a refinery knows the source (field or blend) of the crude oil and its AGI gravity, it can estimate the volume of various products that can be economically refined from the oil. API gravity is also a proxy for the degree of contaminants in the oil, particularly sulfur, since it tends to bind to heavier petroleum molecules.

As a general rule, it is the relative values of petroleum products that set the API gravity differentials on crude oil. For example, a rise in heavy fuel oil prices, relative to gasoline and diesel, will cause the differential to decline. Also a rise in crude oil prices will typically increase the spread between light and heavy petroleum products, thus increasing the API gravity price differential. In the past, posted prices and API gravity differentials were slow to change—new bulletins were issued only a few times a year. However, now that daily oil prices are set in extremely active futures markets, the bulletins are changed at the end of each trading day.

Posted price bulletins provide specific guidance regarding the terms and conditions refiners offer. For example, Chevron's bulletin states the following for oil purchases east of the Rockies:

> Effective 7:00 a.m. on the date(s) indicated, and subject to change without notice and subject to the terms and conditions of its division orders or other contracts, Chevron will pay the following prices ($/bbl) for oil at 40.0° API and above, except as noted below, delivered into pipelines for its account into the custody of its authorized carrier or receiving agent. The posted prices for crude oil transported by truck or barge will be the following prices less appropriate trucking or barge cost. The posted prices for crude oil received offshore will be the following prices less appropriate transportation and quality costs to get the crude on shore.

The posted price system has been the primary basis for determining relative crude oil prices in North America, but it does not set the general level of oil prices. That is now the role of futures exchanges, which are the subject of chapter 4. In recent years, the significance of posted prices has diminished, and they now mostly reflect spot prices reported by the trade press.

The Spot Market

Traders refer to a "one-off" trade, a quick deal made between two parties that have no expectation of a repeat, as a "spot" transaction. In contrast, deals made with the expectation of a continuing relationship are "term" transactions. It should come as no surprise that regular customers at a restaurant tend to get better service and treatment than infrequent diners do. The reward may be a better table, a shorter wait, a free dessert, or even a price discount. Such incentives simply reflect the fact both parties benefit from a dependable relationship. The restaurateur recognizes that he can count on regular customers to fill the tables when times are tough, just as his best customers know they can get a booking during holidays.

Energy markets often reflect a similar behavior to restaurateurs and their customers. Constantly re-contracting is costly, but on the other hand, unexpected fluctuations in demand and supply require a measure of flexibility. Most refiners have a number of long-term purchase contracts or supply sources to ensure a predictable flow to the refinery. They also purchase oil on a short-term basis to balance their needs. In the era dominated by vertically integrated companies, traders referred to the spot market as a "balance wheel." They also called it the "Rotterdam" market, denoting the most important refinery center in Europe and its largest trading center. For three decades following World War

II, the spot oil market was small—estimated at around 5% of total oil flow. Pricing was opaque, and the only reporting was episodic in the context of generalized reports on market conditions.

In the first oil shock, the spot market took on a new role; it began to signal OPEC about how much consumers were willing to pay for crude oil. In the context of a crisis, spot prices soared and official OPEC prices followed soon thereafter. In 1973, high spot prices reflected consumer panic over oil shortages in the Netherlands and the U.S. and the weakening grip of major oil companies on the production and control of OPEC crude oils. In the interim period between the two oil shocks, various OPEC nations solidified control of their own oil resources by expropriation or purchasing the operating companies' assets. The Iranian Revolution accelerated the trend and, by 1980, the only major source of long-term assured supply for the majors was from Saudi Arabia. Even there, however, the ARAMCO partners played a diminished role. Slowly, but certainly, the kingdom acquired their interests, renaming the enterprise, Saudi Aramco.

From 1979 through the early months of 1981, the world oil market had a two-tiered price system. Spot prices would spike upwards; then, OPEC would set a new, higher, official price for the marker. All of the other cartel members would follow suit and adjust their official prices accordingly. Prices of various OPEC crude oils varied substantially as the market was rife with misinformation. Some OPEC members, such as Iran, pursued aggressive pricing strategies, demanding high prices and breaking term contracts. Other countries, particularly Saudi Arabia, sold only at official prices.

Throughout the period, volume on the spot market grew. John Treat, who was the Vice President of NYMEX when it launched its crude oil contract in 1983, estimated that the spot market had grown to over half of the total oil in international trade—a ten-fold increase from its size a few years before. Essentially, the traditional paradigm of the industry was broken. Refiners could not depend on a reliable flow of specific crude oil streams from equity investments or long-term contracts. The issue for refiners was not just pricing; the reduced dependability of long-term contracts disrupted the smooth flow of oil and reduced efficiency, because refiners had to switch back and forth between different types of crude oils. It was obvious that the structure of the business would have to change, but it was unclear how it would evolve. Since the major oil companies could no longer integrate up to control OPEC resources, students of Paul Frankel predicted that OPEC national oil companies would integrate down to own and control refineries in consuming countries. The prediction fell flat; instead, a new and far more flexible market structure emerged, one with futures trading at its center.

Price Indexing

Much of the pricing disarray during the period of the Iranian Revolution reflected extremely poor information. For decades, either the major companies or OPEC had set prices. Prior to OPEC there was some confusion about the real price of oil due to complex tax laws, but once OPEC took charge, the official price of oil was all too clear, reported by scores of journalists covering every detail of a cartel meeting.

On the other hand, the spot market was a classic black box and an even blacker sort of trade. Typically, OPEC members did not want to admit participating, and the prices paid by some oil companies embarrassed them. The market was crowded with middlemen all eager to introduce a bashful refiner to a blushing producer, but the dance was best done in private. One of the first reporters to recognize the significance of the evolving market and to begin to report prices was Marshall Thomas of *Petroleum Intelligence Weekly* (PIW). He described the activity as a "club" and a rather exclusive one. Over time, he was able to gain admittance to the club and gather the confidence of enough traders to track the general trend of spot prices.

Petroleum Argus began publishing systematic crude oil prices in 1979, and in 1981, *Platts Oilgram* also began publishing crude oil prices for international trade. Platts had been reporting petroleum product prices for years, but the only crude oil prices they collected were postings and official cartel prices.

The trade press referred to the crude oil prices they published as "assessments." This was a carefully chosen description reflecting two features of the market. First, in the early years, reporters seldom had access to precise contract information. The reporting proceeded by telephone survey and the price data could not be verified. Frequently, the reported prices were not an actual deal, but rather a deal a trader might have heard about secondhand. In general, traders tended to be tight lipped about their own negotiations, but were willing to report what they thought others were doing. The second problem was the nature of crude oil itself. Petroleum products are standardized; crude oil is not. Some sales blended two or more oil streams together or the cargo might have a slightly different characteristics than usual (e.g., abnormal API gravity or sulfur). All of these nuances affect pricing, so to get a true reflection of market value the reporter would have to make price adjustments.

The difference between a price assessment and a precise contract value was to become significant two decades later concerning electricity and natural gas markets. At the time, however, the key point was not how the trade press collected and published price information, but how the industry used the data. Published trade press product prices had been used for years to "index" contracts between buyer and sellers. As noted, it is costly to renegotiate purchase contracts daily (or even hourly) as market prices change. Consequently, buyers and sellers frequently agree to the type of product, location of delivery, volume, and so on, while leaving actual prices to data published by the trade press at the time of delivery. The majority of oil and natural gas sales contacts in physical markets depend on prices from the trade press in one form or another.

Because price surveys were used to index sales contracts, traders referred to them as "index prices." This turned out to be an unfortunate misnomer. It implied that the indexes were rigorously calculated, something like the Dow Jones Industrial Average and other financial indexes.

The dispersion of prices due to differences in location and quality is normally not very great. Most regional price spikes for oil are short lived, because a variety of transportation options can normally redistribute products in a few days. Electricity and natural gas are different, however, because they have a fixed delivery grid. If there is a bottleneck in the transmission system, it can take months or years to correct. During the California energy crisis, robust economic growth combined with low regional hydroelectric supplies. Natural gas generation was the primary substitute for hydroelectricity, and the sudden surge in demand strained the pipeline delivery system into California. The resulting tight market was ripe for manipulation, and a number of traders responded by reporting prices aimed at enhancing the value of the sales contracts based on published indexes.

When rumors of the manipulation reached regulators, the Federal Energy Regulatory Commission (FERC) and Commodity Futures Trading Commission (CFTC) commenced an investigation as to prices reported and the methodology used. All parties recognized that there are substantial benefits to indexing purchase and sale contracts. Consequently, a deal was hammered out which formalized the price collection and reporting process. Companies now submit data on actual transactions to the trade press, which receives certification from FERC for reporting gas and electricity prices. Certification requires a review of the methodology, annual audits, and other measures to ensure accuracy. FERC does not have regulatory responsibility for crude oil or petroleum product pricing, but the improvements in gas and electricity reporting have also benefited oil price reporting.

Crude Oil Pricing in Today's Market—The Physicals

Today the vast majority of physical contracts index prices to futures exchanges or the trade press. Although this does reduce contracting costs, it sometimes creates another issue in that the number of actual deals being made in the physical market can be a small proportion of total trade volume, or put more colloquially, there is some risk that the tail is wagging the dog. Fortunately, the problem is normally of modest consequence, primarily because most trade press–reported prices are not about futures exchange prices (the primary markers,) but are about all the quality and location adjustments necessary to relate specific geographic delivery points and oil quality variations to the general movement of the market. Typically these adjustments are quite small compared to the broader movements in marker prices and usually do not vary significantly from day to day.[3]

Some crude oil prices quotes are f.o.b. at the country of origin's primary loading port or pipeline junction. Other crude oils price quotes are as delivered c.i.f. prices to specific refinery centers or pipeline junctions in the crude oil's major market. A few oils have price quotes at multiple locations. Argus (2008), for example, assesses Russian Urals crude oil based on either delivery to Northwest Europe (Urals NWE) or delivery to Augusta, Italy (Urals Med). They also assess Urals as a netback to the Baltic at Primorsk, Russia (Urals fob Primorsk) for North West Europe or to the Black Sea at Novorossiysk (Urals fob Novorossiysk) for Italian deliveries. Platts provides assessments of all of the above and adds some additional delivery points. It is important to be clear about the geographic point of price assessment, because transportation costs can make a significant difference.

Timing has become increasingly important in sorting out a crude oil's value. It can take up to 60 days to move crude oil from its point of production to a refinery, and producers have to plan production off-take ahead. Terminals and refineries have storage tanks, but they seldom store more than a few days of oil. As a practical matter, producers usually identify a cargo's destination before production. This means buyers contract for oil days in advance of actual delivery and the designations of *f.o.b.* and *c.i.f.* primarily determine who arranges transportation and bears that cost, not the point where the oil changed hands. In short, almost all oil sales are forward sales; delivery is not immediate, but forthcoming in a matter of days or weeks. Thus, prices struck at market closing on a particular day will vary slightly depending on the anticipated dates of delivery. Historically this has not been significant, but the high market volatility of the last few years has caused the trade press and traders to take account of delivery date differences when assessing the value of a crude oil stream. Traders base these adjustments on the forward "price strip" which is primarily determined in futures markets.[4] Chapter 4 examines the significance of the forward price strip in detail.

Price volatility affects the manner in which buyers and sellers quote prices. Historically, most crude oils were quoted at a flat price without reference to a marker. OPEC's marker price was stable which simplified relative adjustments for quality and location. In 2008, however, futures prices in the NYMEX market varied by as much as $10 per day. Consequently, traders often write sale contracts for specific crude oils as a discount or premium to the closing price (or average of closing prices) in futures exchanges. For example, ANS delivered to the U.S. West Coast is often discounted $2 from the closing prompt price of light sweet crude oil on NYMEX exchange. Alternatively, traders index prices to the trade press, tied to specific future dates.

Although the largest volume of crude oil futures trading is for light sweet crude oil at Cushing on the NYMEX exchange, the most important price source for international crude oils is Brent primarily traded at the ICE in London. This is because of geography—the importance of the European market, the proximity of Russian, Caspian, African, and Middle East producing areas. The North Sea market is also the most complex, because so many of its fields are in decline and because of its early history in developing a complex system of forward trading. The industry often describes the physical market, particularly the North Sea, as "wet" barrels and, likewise, financial trades are "paper" barrels (see chapter 4).

At present, crude oil trading divides into seven major regions – the North Sea, Russian-Caspian, the Mediterranean, West Africa, the Arab Gulf, Asia-Pacific, and the Americas. A few streams fall out of the traditional categories—such as Nile Blend from Sudan and the newly produced crude oils from Sakhalin Island in Eastern Russia. The principal trading centers are New York, London, and Singapore, and the possible emergence of Dubai as a new center. Tokyo has a futures exchange for oil trading, but volume so far has not been significant.

North Sea oil

Early discovered fields, such as Brent and Forties in the North Sea, are now a fraction of former volumes. Consequently, the industry has adjusted physical trading to meet the new reality. Historically Brent crude oil was the key benchmark for spot trading and the North Sea forward market. However, according to both Argus and Platts, the number of free trading cargos of Brent declined to the point that Brent prices may not represent the best valuation of North Sea crude oils. A shift in price reporting occurred in 2002 to reflect the changing circumstances. Table 3–1 describes key North Sea crude oils, their physical qualities, and primary location for pricing.

Table 3–1. Major North Sea crude oils

Crude oil	API gravity	Sulfur	Pricing points
Brent	37.9°	0.45%	Sullom Voe, UK
Forties	41.8°	0.49%	Hound Point, UK
Osenberg	37.8°	0.27%	Sture, Norway
Ekofisk	37.5°	0.23%	Teesside, UK
Flotta	36.9°	0.82%	Flotta, UK
Statfjord	39.1°	0.22%	Statfjord platform

Recall that the North Sea oil market has traded both spot and forward wet barrels since the late 1970s. North Sea forward contracts may settle financially until the oil schedules for delivery. That is, 21 days before the date of delivery (23 days on Friday), the contract becomes a firm commitment to ship the oil. This is in contrast to futures, options, and swaps, which normally settled for cash even after the contract closes. "Dated" North Sea oil is oil past the date when it is committed to delivery. In other words, traders are actually buying and selling the real stuff. Moreover, because dated Brent links closely to the Brent futures market, it indexes to crude oils throughout the globe.[5]

Both Argus and Platts have constructed a "dated" or "dated Brent" price index for North Sea oils slated for delivery in 10 to 21 days. They combine prices and volumes of the first four streams in table 3–1 to create an index, referred to as "BFOE." Both companies apply a similar methodology but base it on different data inputs. Their methodological statements describe the details, and there are several key components worthy of mention. The reporters use reported swaps or contracts for difference (CFD), flat prices, and exchange futures for physical (EFP) to assess the index in the 10 to 21 day window before delivery. The aim is to choose the least expensive of the four BFOE crude oils, or as Platts (2008) states: "the most competitive grade at the margin." The reasoning is that Brent ought to be the least expensive of the four oils, but the low volume of trading may cause occasional distortions; if that happens, one of the three other crude oils will substitute in the index.

Russian-Caspian oil

Russian Urals blend moves by pipeline to central Europe and by tanker to Northwest Europe from Primorsk, Russia on the Baltic Sea and to Southern Europe from Novorossiysk, Russia on the Black Sea. Most assessments are for delivered prices at Rotterdam, the Netherlands or Augusta, Italy (table 3–2).

Table 3–2. Major Russian-Caspian crude oils

Crude oil	API gravity	Sulfur	Pricing points
Urals	31.8°	1.30%	NWE, Italy, Baltic, Black Sea
CPC Blend	43.9°	0.54%	Italy, Black Sea
Azeri Light	35.8°	0.14%	Italy, Black Sea
Druzhba	31.8°	1.30%	Hungary, Slovakia, Poland, Germany, Czech Republic

The CPC Blend of Caspian oil from Russia and Kazakhstan moves by the Caspian Pipeline Consortium to the Black Sea near Novorossiysk and then ships by tanker to Italy. Azeri crude oil from Azerbaijan moves primarily by pipeline to Supa, Georgia on the Black Sea and then by tanker to the Mediterranean.

Deliveries to central Europe move through the old Druzhba Soviet pipeline system, and the trade press assesses prices at various delivery points.

Almost all Russian and Caspian crude oil price as a differential to the Brent futures market or to dated Brent, as published by Argus or Platts. Prices reflect deliveries between 5 to 20 days ahead for Russian oils to Italy and 10 to 21 days ahead for delivery to NW Europe. Caspian oil, the CPC Blend, prices 10 to 30 days ahead of delivery. Pricing for delivery on the Druzhba pipeline is for the balance of the month.

Mediterranean oil

A number of countries along the Mediterranean, Libya, Syria, Algeria, and Egypt supply oil to Europe (table 3–3). These countries set their official sales prices (OSPs) for their contracted oil based on North Sea prices. For example, Egypt set its OSP for September 2008 for Suez Blend at $4.25 per barrel below dated Brent. In addition, oil flows from Northern Iraq and from Baku, Azerbaijan to the Turkish port of Ceyhan on the Mediterranean Coast.

Table 3–3. Major Mediterranean crude oils

Crude oil	API gravity	Sulfur	Pricing points
El Sider	36.2°	0.49%	Libya
Syrian Light	38.0°	0.76%	Syria
Iran Light	33.7°	1.50%	Sidi Kerir, Egypt
Suez Blend	30.4°	1.65%	Egypt
Saharan Blend	46.0°	0.10%	Algeria
Zarzaitine	42.8°	0.06%	Tunisia
Iraq Kirkuk	36.0°		Ceyhan, Turkey

Tankers and the Sumed oil pipeline deliver oil from the Arab Gulf to Sidi Kerir, Egypt—a Mediterranean port. Iran prices many of its spot oil sales at this delivery point. (Most other Arab Gulf crude oils are priced f.o.b. at the country of origin.) Typically the press assesses Mediterranean crude oil 10 to 25 day ahead of delivery, excepting Saharan Blend, which is 15 to 35 days, and Kirkuk, which is 5 to 15 days ahead.

West African oil

Deepwater exploration off Nigeria, Angola, and other West African countries has had a dramatic impact on the volume and types of crude oil produced in this region (table 3–4). So much so that Angola joined OPEC in 2007, the first new country to do so since 1971.[6] West African crude oils are priced as a differential off dated Brent assessments or Brent futures. Some oil is linked directly to trade press assessments. In other cases, the producing country sets an OSP for the coming month tied to trade press North Sea prices. For example, Nigeria's OSP for Bonny Light for October 2008 was set in mid-September at $2.70 above dated Brent. Prices are reported for crude oils 20 to 45 days ahead of delivery.

Table 3–4. Major West African crude oils

Crude oil	API gravity	Sulfur	Pricing points
Bonny Light	34.5°	0.14%	Nigeria
Forcados	30.0°	0.15%	Nigeria
Cabinda	32.5°	0.13%	Angola
Zafiro	29.5°	0.26%	Equatorial Guinea

Mideast Gulf

Although the countries bordering the Arab Gulf produce 30% of the world's crude oil, they follow, rather than lead, market prices. This is primarily because the key countries, such as Saudi Arabia, prefer to market their crude oils through long-term contracts, rather than spot sales. Moreover, most Mideast sales contracts have destination clauses. That is, netback prices vary depending on where the crude oil ships. Another feature of Gulf crude oils is the higher concentration of sulfur, which distinguishes them from African and North Sea crude oils. Dubai and Oman are the only crude oils traded in sufficient daily volume to determine prices (table 3–5). Consequently, most Gulf crude oils prices reflect these prices, but traders, in turn, ultimately rely on the North Sea market as the price guide. According to Argus (2008), the key crude oil is Dubai, and most sour Gulf crude oils price as a differential to it, including Oman. However, Oman has a much larger production volume than Dubai, and it is the marker for the Dubai Mercantile futures exchange, so this may change. Murban and Lower Zakum trade as differential to Dubai. Banoco Arab Medium (produced in Bahrain) is similar in quality to Arabian Medium from Saudi Arabia. Its price sometimes substitutes for the price of Saudi crude oils.

Table 3–5. Major Mideast Gulf crude oils

Crude oil	API gravity	Sulfur	Pricing points
Dubai	31.0°	2.04%	Dubai
Murban	40.4°	0.79%	UAE
Lower Zakum	39.2°	1.10%	UAE
Oman	33.3°	1.06%	Oman
Banici AM	31.8°	2.45%	Bahrain

Gulf crude oils are, of course, some distance from their ultimate markets, so purchases are several months in advance. The press adjusts price assessments for the current month from forward swaps that relate Dubai to Brent futures. Much of the trading for Gulf crude oils is based on the Singapore market, so prices are determined daily on Singapore time.

According to Platts (2008), Saudi Aramco transmits its OSP to its contract buyers once a month for the following month. For Asian buyers, the oil prices on the average of Dubai/Oman assessments with a differential. For European buyers, the crude oil is priced f.o.b. at Sidi Kerir in the Mediterranean or delivered to NW Europe, based on Brent futures. For U.S. buyers, the price is a differential to WTI. Saudi Aramco sells five types of crude oil: Super Light, Extra Light, Arabian Light, Arabian Medium, and Arabian Heavy.

Asia

The Asian crude oil market is the least transparent of all the major markets (table 3–6). This is because there is not one large source of oil in the region; also, the buyers are diverse and spread over great distances. The largest source of oil is in China, but it does not trade, instead going straight to Chinese refineries. Asia's trading center is Singapore, but the only futures exchange trading oil is in Tokyo at the Tokyo Commodity Exchange (TOCOM). The marker for the exchange and basis for settlement is an average of Dubai and Oman oil as reported by the trade press, referred to by TOCOM as Middle East crude oil. Crude oil prices on the exchange are in yen, but converted from dollar values.

Table 3-6. Major Asian crude oils

Crude oil	API gravity	Sulfur	Pricing points
Minas	35.0°	0.08%	Indonesia
Tapis	46.0°	0.02%	Malaysia
Cossack	48.0°	0.04%	Australia
Bach Ho	39.0°	0.04%	Vietnam
Nile Blend	32.8°	0.05%	Sudan
Sokol	37.9°	0.23%	Sakhalin, Russia

Historically, Minas was the largest and most important oil traded in the region. Although its production has declined, Minas remains the marker for other Indonesian crude oils. Indonesia sets reference OSPs for its crude oils, known as Indonesian Crude Prices (ICPs). Production of Tapis, a light Malaysian crude oil, has also declined. It had been widely used as the marker for light sweet Asian crude oils. The trade press reports spot prices for Tapis and a Committee of industrial experts sets the Tapis Asia Petroleum Product Price Index (APPI.)

Actual spot transactions of Tapis have been few in recent years, and in December 2008, Platts began marking it and other Asian crude oils to dated Brent instead.

Most Asian crude oils price with a delivery 15 to 45 days ahead. Australia's crude oils price 30 to 60 days ahead and crude oil from Sakhalin price at the month of loading.

The Americas

The North remains the most important, even though its prices mark few oils outside the hemisphere. The reason is its size, in both financial and physical trading. The U.S. imports crude oils from all over the world, and its New York and Houston traders are legendary. NYMEX dominates price movements, and as the futures market rose to prominence, traders referred to it as the "Merc jerk." Companies like Chevron and ExxonMobil post prices, but they are mainly differentials from NYMEX's daily light sweet crude oil closing price.

Most crude oils imported to the U.S., from the North Sea, Africa, the Arab Gulf, or surrounding countries, are priced to compete with domestic oils. ICE and NYMEX futures often move in slightly different directions due to the time zone change, different refinery structures, and different types of product demand. As noted with respect to Arab Gulf oils, this has resulted in destination pricing for many oil exporters.

Both Canada, to the north, and Mexico, to the south, export crude oils across their borders to the United States. Canada has had a surge of production from its oil sands, identified as a synthetic crude oil in table 3–7. Oil sands bitumen is usually upgraded before export, and it can be refined in conventional refineries. Mexico's export mainstays are Isthmus, which is comparable to many crude oils landed in the Gulf Coast, and Maya, which is sour heavy crude oil and requires sophisticated refining.

Table 3–7. Major American crude oils

Crude oil	API gravity	Sulfur	Pricing points
WTI	40.0°	0.28%	Midcontinent
LLS	37.4°	0.31%	Louisiana
MARS	29.1°	1.83%	Louisiana
ANS	27.5°	1.11%	U.S. West Coast
Midway Sunset	13.0°	1.20%	California
Isthmus	34.0°	1.30%	Mexico
Maya	22.0°	3.30%	Mexico
Synthetic	32.5°	0.13%	Canada
Cusiana	43.0°	0.12%	Columbia
Escalante	24.0°	0.17%	Argentina
Oriente	24.5°	1.30%	Ecuador

The Rocky Mountains divide the U.S. into two primary petroleum zones. The West Coast is less dependent on imported oil because of its production from Alaska and California. California produces mainly heavy oils, with its largest fields being Midway Sunset, Kern River, and Wilmington. Most West Coast refineries depend on ANS crude, which is comparable to imported Oriente and Arabian Medium. Because the refinery feed stock is heavier than normal and there is virtually no demand for heavy fuel oil, the region's refineries are highly sophisticated and manage to convert most of the residuum to diesel or gasoline. During the period when Alaska North Slope was in full swing, the West Coast had a crude oil surplus, with prices well below East-of-Rockies comparables. The surplus lasted over a decade, but now that the North Slope is in decline, the West Coast, once again, requires imports to balance the market. This has tied West Coast oil prices closer to Asian markets; although, given the vast shipping distances the region remains one of the more isolated oil markets.

The mid-continent has extensive onshore production and offshore production from the Gulf Mexico. There are also substantial imports to Louisiana and Texas. The primary terminal capable of handling supertankers for long-haul imported crude oils is the Louisiana Offshore Oil Port, or LOOP for short. A major petroleum product pipeline, the Colonial, ships products north. The East Coast has virtually no production and depends on imported products, imported crude oil for its refineries, and interstate transfers.

As mentioned earlier, the NYMEX's marker is light sweet crude oil produced and delivered in the mid-continent. Historically, crude oils gathered by various pipelines in Texas and Oklahoma classified as West Texas Sour or West Texas Intermediate, and just about all the companies posted for these oils. Posted prices in today's market are more specific. For example, Chevron posts crude oils as diverse as Altamont yellow wax, Heavy Louisiana Sweet, and New Mexico Sour. These oils do not compare easily to the NYMEX marker, and the markets are too thin for the trade press to do an assessment.

North American crude oil prices depend on the month of delivery (quoted spot prices normally reflect delivery the following month). Waterborne imported crude oil prices 10 to 45 days ahead. ANS prices at the point of delivery for the month of arrival.

Physicals and Futures

When discussing the complexities of trading all the various grades of crude oil at points all over the globe, it is easy to lose sight of the defined linkage between physical and futures trading. As is discussed further in chapter 7, the oil industry is one of continuous flow. Inventories balance the flow, but with the exception of strategic reserves, they are small relative to the total daily consumption. Even a minor change in inventories can have a significant impact on oil prices, immediately expressed in the futures market.

A change in futures prices flashes to the physical markets by the end of the trading day. Generally, a rise of futures prices provokes a similar rise of all crude oil prices in physical markets, with a series of relative adjustments reflecting difference in quality or the circumstances peculiar to a region. The change in crude oil prices shifts planning for refineries, and affects petroleum product prices. The new price signals in wholesale and retail markets ripple through the vast number of marketers, commercial users, and retail customers who make decisions to buy, store or use oil. These decisions, in turn, feed back to refiners who seek to buy more or less crude oil, depending on the signal. The resulting shifts in demand for crude oil manifest themselves in shifting demand for NYMEX's and ICE's marker oils, where the signal gets conveyed back from physical markets to commodity traders who bid the price of futures up or down. It is a vast circle of interdependence.

Notes

1. For example, a producer may minimize the risk of a price collapse by selling crude oil futures contracts. If the spot price is lower at the time the crude oil actually flows, then the producer makes a profit on the futures contract, but receives less for the physical sale of the product. Of course, the reverse can happen too. If spot prices at the time of production are higher than the price in the futures contract, the producer takes a loss on the financial contract, but makes extra on the physical sale of the oil.
2. The term "prompt price" refers to the price determined for the first delivery month of futures trading.
3. The efficiency of crude oil pricing depends more than ever on prices set in futures market trading. In 2008, there were allegations that hedge funds and other financial traders had manipulated this market. This issue is addressed in chapter 4.
4. Forward price strips are the series of price for each future month of delivery as determined in a futures exchange or in the over-the-counter market. Rising prices over future periods are described as "contango," falling prices are in "backwardation" (See chapter 4).
5. North Sea physical markets trade on a weekly cycle while Brent futures trade on a monthly basis. This results in some complications and an imperfect relationship between physical and futures prices.
6. Ecuador rejoined in 2007 after an absence of 15 years.

4

ENERGY COMMODITY MARKETS

How NYMEX Won the Lottery

Accidents are often at the root of history. However improbable it seems, energy commodity markets developed from the convergence of two unlikely events. In 1978, the NYMEX futures market was about ready to fold its tent, facing stiff competition from the Chicago exchanges for its agricultural products. The exchange was desperate to find a product all its own, but it faced some major constraints. Two recent contracts dealing with potatoes had failed, thus development time and the CFTC prohibited NYMEX from introducing a wholly new product. In the 1970s, the exchange had introduced a Rotterdam-based oil futures contract that failed, but it meant that the exchange already had approval to trade oil (Falloon 1995). This time, NYMEX chose heating oil delivered in New York harbor, and opened the contract in November 1978, just at the point when oil workers revved up their strike in the Iranian oil fields. Anxiety over oil supplies, soaring prices, and OPEC's breakaway from the majors did the trick and the heating oil contract proved highly successful.

Despite the extraordinary success of energy futures markets at NYMEX and ICE, energy futures are not the most widely traded contracts. Futures and options trading in equities and interest rates now set the pace. Figure 4–1 describes data from the Futures Industry Association (FIA) on the volume of futures and options contracts. Over sixteen billion contracts traded in 2008, but only 3.3% of these related to energy. Metals and agricultural products totaled 7.0%; double oil and natural gas. It is important to point out, however, that these figures are for the contract volume, not value. When oil sold for $145 per barrel, a single contract was worth $145,000. Based on value, energy's share would have been much higher.

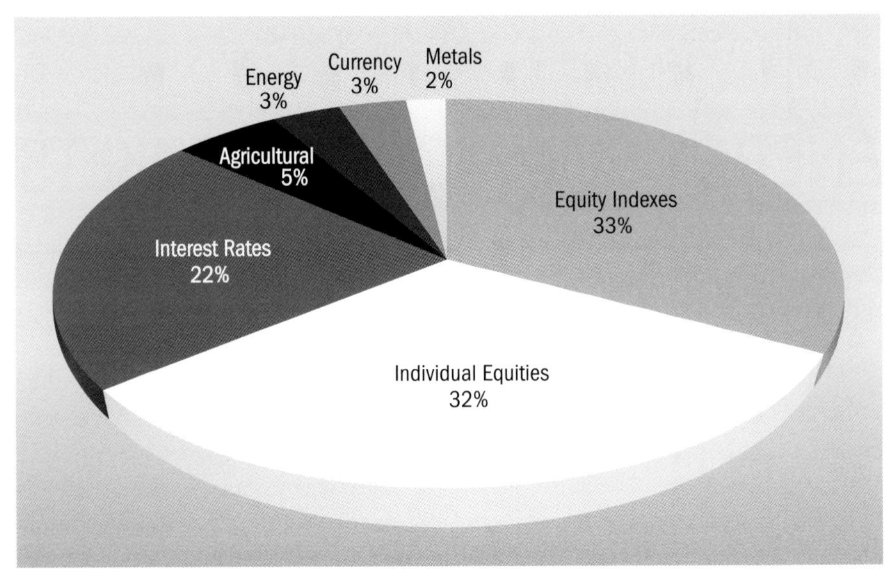

Fig. 4–1. Exchange volume of futures and options contracts 2008

Today's system of energy trading grew out of existing trading arrangements from other commodity markets (mainly agriculture and metals) and from the financial industry. Present energy markets are highly sophisticated and complex, although most of the activities and concepts have roots back to the nineteenth century.

In the eighteenth and early nineteenth centuries, there were a variety of agricultural exchanges throughout Europe and North America. These exchanges were a cash auction where farmers brought their products to a central trading area, and millers and other merchants would bid for the grains. The development of telegraph and telephone communications, however, made most of these exchanges redundant. Farmers could rely on prices set in Chicago or London and contract directly with millers or other marketers at these prices. The last active cash agricultural exchange in the U.S. was in Portland, Oregon for white wheat. This type of wheat is unique to the Pacific Northwest and sufficiently different from the red wheat traded in mid-continent that it requires its own unique pricing. The exchange, however, proved too costly to be justified and closed its floor in the early 1980s, replaced by bilateral negotiations mostly over the telephone.

As regional agricultural exchanges consolidated with improved communication, they developed forward trading, which evolved into futures trading. In the Northern Hemisphere, strong seasonal differences produced an annual pattern of crop production significantly different from consumption. Wheat, corn, and other grain products had to be stored through three seasons before the summer harvest replenished stores. Moreover, farmers had to decide the type and amount of each crop well in advance of the harvest. To make rational decisions they needed price guidance. Farmers could contract directly with millers, packers, and so on, or join a cooperative, but there was huge uncertainty, substantial counterparty risk, and poor price information.

The Portland wheat exchange, like all cash commodity exchanges, was unregulated. Likewise, trading of futures-like contracts in the bilateral market has so far escaped formal oversight. In contrast, the CFTC regulates futures exchanges. Regulatory oversight also became a significant issue in the design and operation of the plethora of electronic energy exchanges that emerged in the 1990s.[1] The issue surfaced again in 2008 concerning the large amount of OTC trading in unregulated financial contracts.

The CFTC describes its regulatory responsibility as follows: "Today, the CFTC assures the economic utility of the futures markets by encouraging their competitiveness and efficiency, protecting market participants against fraud, manipulation, and abusive trading practices, and by ensuring the financial integrity of the clearing process." (CFTC Web site) To achieve this goal, the CFTC collects data on those that participate in futures trading and defines two categories of activity—commercial and non-commercial participants. Commercial participants are those that are active in the industry, such as oil and gas producers, refiners, energy transporters, and so on. The CFTC assumes that the principal activity of these traders is hedging. Non-commercial traders are hedge funds, investment banks, and so on that have no direct need to hedge oil and gas. Analysts presume this group be "speculators." The categorization is obviously imperfect, and may be refined in the near future. The CFTC releases data on the relative positions of the two groups once a week.

The Nature of Derivatives

Most people refer to futures exchanges as commodity exchanges, but their purpose is not to match buyers and sellers of actual commodities. These exchanges trade "derivatives." John Hull (2009 p. 1) explains derivatives as follows: "A derivative can be defined as a financial instrument whose value depends on (or derives from) the values of other, more basic, underlying variables. Very often the variable underlying derivatives are the prices of traded assets." Put tersely, futures traders are trading the right to buy or sell oil or natural gas, not the commodity itself. There are, of course, all sorts of stories about futures traders that forgot to close out a contract and end up with a thousand bushels of wheat (or hog bellies) delivered to their backyard. These are largely myths, since the normal point of delivery is a commercial site. Nonetheless, the failure to close out a futures contract can prove to be huge hassle if it's not intended. Futures markets depend on the fact that the underlying asset can be delivered if necessary, and exchanges have a standardized process for converting futures to physicals. Although there must be a linkage, traders clearly distinguish the two activities, commonly referring to contracts traded on a futures exchange as "financials" and the trade in the commodity itself as "physicals."

In energy markets, there is frequently parallel trading in physicals and derivatives. A contract for physicals buys or sells the commodity, and a derivative contract manages price risk. These contracts trade through futures exchanges or independently in the OTC or bilateral market. Bilateral trades in derivatives may transform to physical deliveries and physical contracts may close out financially, depending on the particulars of the deal and the interests of the parties.

Physical sales contracts usually involve just two parties, the buyer and the seller. For example, consider a refiner seeking to buy oil delivered in six months at a price set today. The refiner matches with a producer who agrees to sell the oil and they agree on the forward price, say $50 per barrel. Six months hence, delivery takes place (usually as a daily flow over a one-month period), the seller invoices, and the buyer pays 20 to 30 days later, depending on the contract. In this type of contract, the parties have fixed the quantity, price, and timing of the forward delivery.

More often than not, however, buyers and the sellers have different ideas about what forward prices ought to be or how much price risk they wish to absorb. That easily blocks agreement. Consequently, most contracts have flexible prices that vary with market conditions. Historically, crude oil traders in North America pegged contract prices to posted price bulletins or to the official sales price (OSP) of the exporting country. Both parties to the deal understood that prices would change with market conditions. Today, traders index most purchase contracts to prices published by the trade press or determined in futures exchanges.

Spot purchases or sales contracts with flexible prices cover delivery terms, such as timing, volume, quality of oil, and so forth, but open up both parties to price risk. Consequently, either party or both may enter into separate price hedging or financial contracts. In the OTC, traders refer to such contracts as "swaps" or "contracts for differences" (CFDs)[2] (Mirant Web site). Although they are fashioned less formally, swaps are similar in effect to futures trading and can be a substitute.[3] Banks, financial services companies, or energy marketers prepared to accept price risk, are usually the counterparty to swaps.

Commodity prices, not physical deliveries, are the issue in financial contracts. Essentially, the parties agree to a cash receipt or payment in the event spot prices change at the time of settlement. The calculation takes the difference between the spot or cash price on the stated date of settlement as compared to the fixed price in the contract. For example, consider a refiner with a purchase contract at index prices. The refiner, as hedger, could enter into a swap with a financial company or marketer. Assume the swap has a fixed price of $50 per barrel. If the spot price turns out to be $40, the refiner would compensate the swap seller by $10, plus commission. If, on the other hand, the spot price turned out to be $65, the swap seller would compensate the refiner by $15, less commission. Instead of just contracting for the physical purchase of the oil at an unknown price six months away, the refiner has locked in an effective price of $50. If the market skyrockets in the interim, the oil refiner will receive enough cash from a derivatives contract to ensure the physical purchase, no matter what the spot price turns out to be. Likewise, if the price of oil plummets, the refiner will get a good deal buying the oil, but will have to pay off the derivatives contract. In either case, the forward purchase price *nets* to a flat $50 (plus brokerage fees).

The motives for financial trading are hedging, speculation, or arbitrage. The main motivation for hedging is to reduce risk. The motives for speculation are less well defined, but the classic example is an investor who has become convinced that the price of the commodity in question will either rise or fall. Speculators are willing to take the gamble and reap the profits if they are right or suffer the loss if they are wrong. In the case of swaps, financial firms often minimize risk by balancing deals with both buyers and sellers. In other words, a financial company can sell CFDs to producers, with a payout if the price goes up. In parallel fashion, it can sell CFDs to consumers, with payout if the price goes down. If the two sides offset one another, the company will collect brokerage fees with little risk, turning speculation into arbitrage.

Traders (and their overseers) pay close attention to their "position"—the net consequence of offsetting purchase and sale contracts. If sales obligations exceed purchase obligations, they are "short." If reversed, if they have agreed to buy more than they have sold, they are "long." A price drop gives the advantage to traders shorting the market and traders who are long see advantage in rising prices.

The Benefits of Parallel Trading in Financials and Physicals

One question is likely to arise early in this discussion. Why bother with swaps or futures; if parties want to reduce price risk, why not simply enter into a long-term forward contract with fixed prices?[4] There are three primary reasons this is difficult to achieve. First, it is often hard to get a precise match between a buyer and seller of all the necessary terms. In the case of the crude oil market, they would have to agree on the quality of the crude oil, delivery locations, volume, timing of deliveries, transportation systems and costs, credit terms, and, of course, the price. In most cases, the cycles of crude oil development (three to ten years) and the cycle of refinery operations (about three months with four seasons) simply don't match. In short, both buyers and sellers prefer a measure of flexibility. Short-term sales or contracts with floating or index prices provide the flexibility but open up both sides to price risk.

The second problem arises from the extreme price volatility in energy markets. If contract prices diverge significantly from spot prices, the odds are good that the disadvantaged party will seek to void the contract. When North American spot natural gas prices plummeted in the 1980s, the entire industry faced a crisis that led to its complete restructuring. As Arlon Tussing famously observed at the time, the most common phrase in the industry was: "I won't take and I won't pay, so sue me." Swaps do not require the singularity of commitment as do long-term contracts in order to achieve hedging goals. For example, an Alaska oil producer can hedge most price swings using almost any marker crude oil, such as sweet light oil at Cushing or even Brent in the UK. Moreover, swaps allow a wide range of diversified counterparties. That is, the producer can hedge with as many counterparties as deemed prudent in a series of smaller contracts, significantly reducing risk. This is much more difficult to do with a particular crude oil stream bound for a specific destination.

Finally, most swaps can be closed out at any time, without disruption in the physical market. For example, if an oil producer has greater than expected field decline, it is relatively easy to get out of a financial contract. In contrast, terminating a contract that involves physical delivery can be far more complicated. Changes in physical deliveries require changes in transportation, new schedules, and so on. In sum, swaps combined with spot or price-indexed sales are more flexible, have lower risk, and have lower transaction costs than physical contracts that roll everything together.

How an Exchange Adds Value

It is clearly not necessary to have an exchange in order to trade CFDs or swaps. It is, however, far more difficult. In order for a CFD to be credible, both parties must have confidence in the source and reliability of pricing. The primary contribution of a futures exchange is transparent price discovery. The Executive Vice President of NYMEX (Wolkoff 2000) describes the role of a futures market in this way:

> Futures markets provide two important economic functions: price transparency (price discovery) and risk shifting (risk management).
>
> Price transparency is the constant reporting to the world of the prices of actual trades being made at the exchange. With tens of thousands of energy contracts traded daily, each reflecting a binding commitment to make or take delivery of a specific commodity, price information is made available in real time, on a virtually continuous basis. Thus the true world reference price referred to earlier.

> Risk shifting, in the secure liquid markets that the New York Mercantile Exchange provides, allows commercial interests to "hedge" the risk of price fluctuations that could affect profitability and planning of their business operations. For the commercial participant, the result is a form of insurance against the financial adversity that can result from volatile energy prices.

A futures contract is an agreement between two parties for delivery of a particular commodity at a specific time, place, and price. Once initiated, a futures contract obligation can be satisfied by taking an offsetting position or by going through the delivery process and taking possession or making delivery of the commodity. The vast majority of market participants opt for the former, making futures contracts one of the most common and useful financial tools.

In 2008, about 246 billion "paper" barrels of crude oil traded on NYMEX and ICE in just three contracts related to light sweet crude oils at Cushing, Oklahoma and North Sea Brent crude oil. This was approximately eight times greater than annual world oil production.[5] Obviously, there was not enough oil at Cushing, Oklahoma or at Sullom Voe to fill even the tiniest portion of this trade. Nonetheless, as chapter 3 described, prices determined in these exchanges effectively set crude oil prices around the globe.

Price transparency

When confronted with futures exchange volume for the first time almost everyone has the same reaction. Why is trading volume so high? The prime advantage of such a large turnover is price transparency and market liquidity. These days an investor would almost have to be blind and deaf to avoid knowing the daily price of oil. Before NYMEX, however, finding out the price of a barrel of crude oil was challenging and connoted a substantial amount of uncertainty. There were anecdotal reports of spot prices, but longer-term prices were not available from any source. This led to inefficient decisions in physical markets.

The high turnover in crude oil and natural gas markets futures contracts is typical of commodities traded in this manner, because as futures trading matures, the cost of trading declines.[6] It then becomes economic to adjust portfolios in response to changes in weather, inventories, or any of the other variables that affect demand and supply.

Trading liquidity

The fact that there are three primary standardized futures contracts for crude oil is significant. The limited number of contracts plus a high rate of turnover is what allows a liquid market. The concept of market liquidity is not as straight forward, as it might seem. Ruben Lee (1998, pp. 50–51) discusses all of its implications:

> One conception of a perfectly liquid market is that of a forum in where it is possible to buy and sell an infinite amount of the asset being traded, at the same time, without delay, and at the same price. As is self-evident, this definition is composed of several distinct components. The difference between the bid and the ask price for small-sized orders, or the '*width*' of the spread, is one measure of liquidity. Another is the '*depth*' of the market, a gauge of the manner in which the spread widens or narrows as the size of the transaction becomes larger. A further element of liquidity is the market's '*immediacy*', namely whether there will immediately be a price at which it is possible to execute all trades. Another aspect of a market related to liquidity is the speed at which orders can be executed, or analogously the expected time market participants need to wait before an order on the other side of the market appears. Yet a further element of the liquidity of a market relates to the dynamic properties of transaction prices, and in particular the extent to which transaction prices diverge from and revert to equilibrium prices, or the so-called '*resiliency*' of the market. [Emphasis added.]

As Lee's expansive description suggests, liquidity and trading costs are clearly related. In commodity and securities markets, they are opposite ends of a scale. In other words, an illiquid market has high trading costs and a liquid one has low costs.

The degree of liquidity can be measured by the difference between bid and ask prices, a cost that must be borne by one or both parties. The housing market is a good example of illiquidity. Frequently, homeowners will complain that it takes months to sell a home. That is not exactly true; houses sell quickly if the seller is willing to discount the price. Instead, the seller is seeking to get the best possible price by a willingness to hold onto the house until enough prospective buyers have seen it to establish reasonable market value.

The cost of trading oil in both the futures and the physical markets is now remarkably low. Physical trading takes some time to put deals together, but the futures market is, literally, immediate. Open outcry trading on the NYMEX runs from 9 a.m. to 2:30 p.m. on normal business days, but with the exception of holidays and weekends, electronic trading is available almost continuously. Most electronic trading platforms list the most recently closed price and the number of bids and offers at various price levels. A trader can purchase or sell large quantities of oil or gas with only the click of a few buttons; execution time is a matter of seconds.

The liquidity of oil, natural gas, and some electricity markets may not have reduced price volatility, but it has reduced the incentive for unnecessary hoarding that is often associated with physical shortages. John Maynard Keynes identified two motives for stockpiling—"precautionary" and "speculative."

When consumers face threats of physical shortages, they have a strong motive to buy more than they need, rather than run out. In the 1970s, fear of gasoline shortages actually exacerbated the problem because consumers chose to drive on the top half of their gasoline tanks, rather than the bottom half. Likewise, speculators, sensing that crude oil prices were on the rise, had a motive to accumulate unnecessary inventories. The rapid rise in oil prices in 1979 during the Iranian Revolution caused speculators to store oil in unused tankers.[7] Similarly, price controls in China and India in 2008 created an illiquid market with fear of shortages in those countries. This provoked unnecessary stockpiling, leading to price spikes and ultimately to the market collapse. Liquid futures and physical markets reduce the incentives to hoard. In the first place, oil will always be available at the market price. Secondly, if traders expect the price of oil to rise there is no need to stockpile, they can simply use the futures market to buy forward.

The rapid expansion of oil trading in 2008 led to accusations that speculation caused high prices. Unlike 1979, however, there was little or no stockpiling outside of China and India. Attempts to manipulate markets are not effective unless traders withhold supplies. In the oil industry, withholding can occur in one of two ways: First, companies and individuals accumulate inventories in order to take advantage of expected higher prices. More seriously, producers may cut back production, stockpiling rather than producing reserves. Through the first half of 2008, OPEC was said to be close to full production capacity, but such data are only estimates and not very good ones. In the 1970s, OPEC's oil supply was said to be "backward bending." That is, as oil prices went up, producers had such a robust cash flow they had no incentive to increase production. It is likely that there were elements of this behavior in 2008, in part, because a large oil surplus quickly emerged in the second half of the year as prices fell. Such behavior, however, is due to rapidly rising prices in the physical market; the futures market reflects the physical market, it does not lead it.

Professor Ronald Ripple investigated the activities of commercial and noncommercial interests during the period of price increase in 2008 and reached a similar conclusion. He found that 53% of the increase in long positions (those that would benefit from rising prices) were due to trading by the commercial group; the group that is primarily hedging, not speculating. "Moreover, of the 44% of the long positions' growth attributable to non-commercials, 20% represents one side of an offsetting calendar spread position." Professor Ripple concluded that commercial, rather than non-commercial traders had led the market up (Ripple 2008).

In assessing the impact of speculation it is important to recognize that oil is not stockpiled to the same extent as many agricultural products or gold. These commodities can be more cheaply stored. Crude oil is refined and sold within a few weeks of its production, making speculation costly and difficult. Oil's "continuous flow" has an important impact on industry's structure, and that characteristic is analyzed in chapter 7.

Reduced counterparty risk

Exchanges evolved from coffee houses and initially were quite informal. Brokers soon discovered, however, that to be successful they had to have trading rules and a defined structure. The starting point is a simple and powerful observation—no one wants to trade in a rigged market. The first step toward creating a fair trading environment is to separate the interests of the exchange from the buyers and sellers that use it. In other words, an exchange does not take a market position. It provides financial guarantees for traded contracts, and is the counterparty to all trades, but exchanges do not gain or lose as prices change because, they match each purchase with a sale. They minimize the risk of default by requiring participants to deposit a substantial portion of the market value of each contract. Each day, a trader's position is "marked to market" through the exchange's clearing house. This means that contracts are valued at current prices rather historic cost or value. If a trader's funds are inadequate, the exchange makes margin calls, and if not covered, the position is closed out. Using this process, an exchange essentially eliminates counterparty risk, a reoccurring problem in bilateral contracts.

Risk shifting

The classic argument to justify the derivatives market is that it allows companies and individuals to shift risk to those better prepared to handle it. Essentially, risk shifting allows diversification and tapping into much deeper financial reserves. A flat price sales contract in the physical market depends on a single counter party's performance and obviously limits diversification. The oil and gas industries have large economies of scale, and this frequently constrains the number of potential counterparties. For example, on the West Coast of the U.S. there are only a few refineries short of crude oil. Given that both crude oil and petroleum product prices tend to move together, refiners do not find flat priced contracts for longer than a month attractive. There are alternatives for Alaska and California oil producers, but they entail high shipping costs. By separating financial and physical contracts, the number and types of counterparties for managing price risk greatly increase. With price issues set aside, it is easier to optimize physical flows, matching the right crude oil stream to the right refinery configuration. The dispute in the winter of 2008–2009 between Russia, Ukraine, and the European community over natural gas is a good example of the disruptions that can arise when parties intertwine financial and physical negotiations.[8]

Entities that deal in swaps usually do not confine themselves to a single industry. Investment banks, for example, trade derivatives linked to securities, interest rates, and a variety of commodities. This provides a large measure of diversification as well as a huge pool of financial reserves. It is interesting to note that in the 2008 financial crisis, the extreme volatility of oil and natural gas prices had little impact on the solvency of banks and other intermediaries that offer energy swaps. In contrast, subprime home mortgages and credit default swaps (CDSs) did have a dreadful impact on bank solvency in 2008. The difference is clear: commodity swaps have a core of buyers and sellers, so banks can perform arbitrage by balancing both sides. Credit swaps, however, are akin to insuring a natural disaster. Homeowners want protection from fire and floods, but almost no one benefits from such events, so those that offer insurance must set aside a large pool of financial reserves if a payout is required. In order to avoid insurance regulation AGI and other offered CDSs in the derivatives market; consequently, the issuing company did not have to amass much in the way of financial backup. So, when the disaster occurred, the insurer skipped town.

How Rules and Market Structure Impact Trading

There are other interesting observations concerning early exchanges. The biggest problem is what economists label the "free rider" problem. Traders have a natural inclination to keep their own trades secret, but wish to know what everyone else is doing. Obviously, that circle does not square; without an organized trading structure, the result is often an incomplete or distorted view of the market. Exchanges solved this problem through open outcry trading. They achieved secrecy for companies and traders that wished to remain anonymous by putting brokers, rather than principals, on the exchange floor. Brokers in the trading pit had to cry out offers in front of everyone. The shouting added a level of frenetic energy to the trading, but it also conveyed objective and complete information. Exchanges required that all trades be recorded, forbidding secret deals. Otherwise, brokers would use information provided by open outcry, without giving the exchange the revenue it needed to cover operating costs. Another feature that provided discipline was the property rights to seats on the exchange. To trade, brokers needed to buy seats, and once they had put up the cash, they then had an incentive to see the exchange succeed. If the exchange failed, the value of their seat would fall to zero.

Much has changed since the exchanges originated. Today, most are publicly owned corporations, and utilize electronic rather than open outcry trading (or some combination). The incentives to free ride the exchange have been greatly reduced because volumes are much higher and brokerage fees have fallen along with the new technologies. In the last year, increased counterparty risk in bilateral markets has greatly enhanced the value of exchange trading. Nonetheless, it is worth remembering a central point: exchanges are like nightclubs, no one wants to go to an empty one. Put another way, exchanges are natural monopolies. Once established, traders will gravitate to arenas in which trading volume is high; the reason is simple, liquidity enhances trading value.

Criteria for Successful Futures Trading

Dennis Carlton (1984) studied the evolution of futures markets noting that until 1970 these markets were limited to metals, industrial products, oilseed, livestock, food, and grain. During the 50-year period before financial products were introduced, the number of contracts increased slightly from 50 to 78. Since then, the number has increased dramatically to include almost every type of financial product, including stock indexes, commodity indexes, and shorts. There is even a traded index capturing stock market volatility.

According to Carlton (1984 p. 242), successful contracts have five features:

1) There is a high level of uncertainty about price and availability of the underlying commodity.
2) There are price correlations among similar products, i.e., basis risk is predictable.
3) There are a large number of participants and an appropriate industrial structure.
4) Transactions have large values.
5) Prices are freely determined without regulation.

The third point concerning industrial structure is particularly significant to the energy industries, because Carlton claims that vertical integration can stifle the development of a futures market.

Black (1986 p. 6) surveyed the literature on new futures contracts and identified six requirements for product success:

1) The commodity must be durable and storable.
2) Units must be homogenous.
3) The commodity must be subject to frequent price fluctuations with wide amplitude.
4) Supply and demand must be large.
5) Supply must flow naturally to a market [(a) competitive cash market and (b) low delivery costs].
6) There must be breakdowns in an existing pattern of forward contracting.

Black discusses each of the above points, and two issues merit elaboration. First, the storability of a commodity is no longer essential. This is because, as Black (1986 p. 7) recognized, "...futures contracts had a forward pricing function separate from their inventory guiding role.... Goods for futures delivery did not have to currently be in inventory as long as they would become available through production." This opened up the potential market for futures contracts to a wide variety of commodities, including crude oil and petroleum products, where the flexible surface inventory is only a few days' supply. Second, the issue of product homogeneity is crucially important to the success of a futures contract. That is because contracts are standardized around a specific product and location. It is not necessary that the commodity be homogenous for a futures contract to succeed, as long as the value between various grades can be objectively determined and location cost differentials are known and stable.

According to Black (1986), success or failure in designing new futures contracts rested on two issues—the choice of the commodity and contract design. Her comments on the commodity itself were particularly interesting. She developed a generalized theory of contract success and used statistical analysis to support it. Her statistical analysis suggested that success or failure of a futures contract rested on the opportunities for "cross hedging." If a new contract concerned a commodity that had a close substitute, hedgers might choose to trade the substitute, rather than the new contract. In this instance, hedgers faced a trade-off between two types of costs. The new contract might be specific for the product they wished to hedge, but if trading volume was low, the spread between bid and ask prices could be very high and shift unpredictably, increasing the liquidity penalty. Alternatively, prices in the established contract might not be perfectly correlated with the commodity for which they sought to reduce risk. The most cost-effective choice for reducing risk would determine success or failure of the new futures contract.

How Options and Futures Contracts Relate

Options to buy or sell forward are valuable tools for reducing risk, but they work best in a transparent and liquid market. Consequently, they normally follow the development of successful futures trading. The right to buy at a fixed (strike) price sometime in the future is a "call," and the right to sell a "put." Even though formal options contracts depend on an established futures market, the idea that underlies such trading permeates many markets. Some "wholesale" outlets charge an annual fee, which gives the member the right to buy discounted goods, but not the obligation. The real estate industry uses purchase options, where the buyer puts money down in order to establish the price at a future time of closure, but is not obligated to complete the deal. Many sports season tickets include the right to buy tickets to playoffs or tournaments, but again there is no obligation to do so.

Conceptually, options are like insurance (more so than futures). A company or person that buys an option to hedge should view it as an ongoing fee or cost to avoid large losses. On the other side of the equation, traders willing to sell options are able to generate revenue without necessarily selling the asset.

In most cases, options expire "out-of-the-money." Because the commodity price is above the strike price for a put or below the strike price for a call, the buyer does not opt to exercise the option. Logically it would seem that if the price of crude oil is $50 per barrel, the right to buy it six months in the future at a strike price of $40 should be worth $10. Traders refer to this as the option's "intrinsic" value. In fact, such a call option is usually worth a lot more. This is because prices fluctuate. The value of an option must reflect both the difference between the strike price and current price of the commodity, as well as the probability that prices will move advantageously before the option expires. The closer a call option contract gets to its expiration date the lower the probability that the commodity price will go higher and therefore the lower the premium value of the option (value above intrinsic value). Traders estimate options values based on the pioneering work of Professors Merton, Black, and Scholes. John Hull provides details on such calculations in *Options, Futures, and other Derivatives*. The key point here is that options are an important part of a future exchange's activities and important tool for risk management. Of the 16 billion contracts traded on futures exchanges in 2008, about half were options.

The Term Structure of Energy Pricing

The concepts underlying spot and term prices, discussed in chapter 3, are essential to understanding the modern commodity market, in which the relationships between short-run and long-run prices are explicit. Futures exchanges offer trading in a series of forthcoming months with individual contracts for each month. The result is in an individual price for each of the forward months. In contrast, the bilateral or OTC market is far more flexible. In these agreements, physical or financial contracts may be offered at flat prices for various periods or at floating prices indexed to publications or an exchange, or some combination—there are no preset rules; contracts do not have to be standardized.

Futures exchange prices are forward prices (but forward prices are not necessarily determined on a futures exchange). The relationship between a series of forward prices constitutes the term structure; traders also refer to a set of forward or futures prices as the forward price strip. As mentioned earlier, the price given for the first month of a price strip is the prompt price. As a practical matter, this corresponds to the spot price.

Figure 4–2 compares natural gas and oil price forward strips from the NYMEX exchange settled on January 2, 2009. Determinants of the shape of a forward strip (whether prices rise or fall) are complex. Generally, forward prices reflect expected future spot prices. That is, when entering into a long-term agreement, neither the buyer nor seller will agree to a price that is much different from their expectation of what they could, on average, buy or sell the commodity for in future periods. The link between the price in the first month to prices in the remaining strip depends on time value of money, the cost of storing a commodity, expected future production costs, and a generalized catch-all referred to as "convenience yield" (the idea that there are benefits associated with having a physical inventory). Gold has low storage costs, so spot prices are seldom much different from forward prices. Agricultural prices usually have a seasonal cycle reflecting expectations about crop yields, the cost of storage, and the use pattern. The natural gas forward price strip has a strong seasonal component, reflecting high storage costs, fixed production capacity, and strong winter demand. There is also a seasonal price cycle for gasoline and heating oil, but the two more or less wash out so crude oil does not usually have a seasonal component.

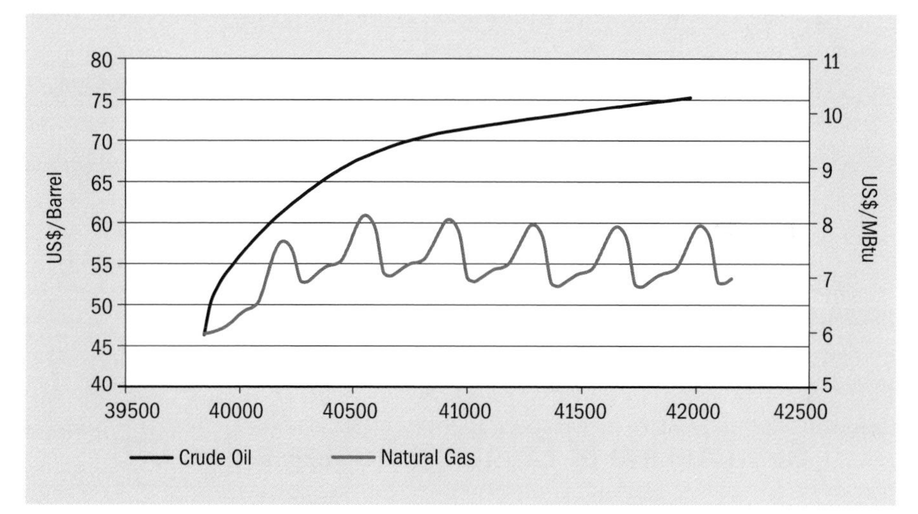

Fig. 4–2. Comparison of crude oil and natural gas forward prices

Figure 4–3 illustrates the forward price curves of oil traded on the NYMEX in August 2003 and January 2009. When forward prices are progressively lower in future months, traders refer to such circumstances as backwardation. This pricing structure might, for example, arise before harvest time, when the forthcoming yield is known to be high and prices can be expected to decline once the harvest is underway. The alternative, contango, refers to market conditions where forward prices are progressively higher than spot prices. In 2003, crude oil prices exhibited strong backwardation, reflecting an expected price decline, traders expected Iraqi oil production to come back online in the near future. Of course, it did not work out that way; prices rose continuously up to the summer of 2008. This price rise was unusual in that both prompt and far-forward prices went up, reflecting strong demand, particularly from emerging markets, and pessimism about long-term oil supplies.

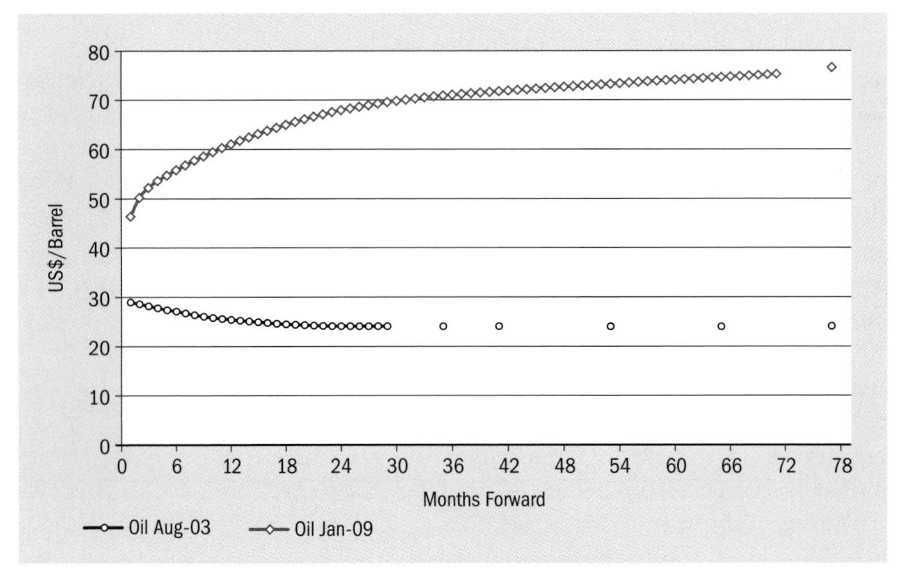

Fig. 4–3. Backwardation and contango in oil price forward strips

Since the price collapse in the last half of 2008, near-term prices have declined much more than those set for four or five years in the future. This reflects the short-run impact of the recession on oil demand and a residual pessimism about longer-term supplies.

Relationship of Exchange Trading to the OTC

Commodity and securities markets operate in continuous trade; i.e., during open hours, assets trade more or less instantaneously, which, as explained, creates a liquid market. These contracts change hands minute to minute and yet no one is concerned that in one hour there might be only sellers and in another hour only buyers. Exchanges provide a class of brokers, known as "market makers." Market makers are willing to hold inventory in order to even out the flow of buyers and sellers. These companies are crucial in providing liquidity. In continuous trading exchanges, prices change throughout the day, with special attention paid to the closing price. Closing prices are widely reported and often linked to contracts. In contrast, the OTC has no special hours, and negotiations often take considerable time to complete.

It is difficult to obtain reliable data on the volume and value of OTC trading in energy markets. Like futures trading, however, it is not significant relative to the number and amount of total derivatives traded. The Bank of International Settlements (BIS 2008) found that there were $684 trillion worth of derivatives contracts in the OTC market in mid-2008, but only $13 trillion were for commodities, and energy is likely to be only about one-half of those.

OTC trading is mainly about price differences arising from regional or quality differences. For example, traders might hedge the price difference between sweet and sour crude oil or between Arabian Light crude oil delivered to California or the Gulf Coast. Basis describes the difference between the price of one commodity slightly different from the next, or in a different location. Basis risk arises when the commodity in question deviates in unpredictable ways from an established relationship. As explained in chapter 3, crude oil varies enormously in quality and location and there are many short-term regional disruptions. Thus, there is often significant basis risk.

Two decades of experience in energy markets suggests that futures exchanges and OTC trading are mutually dependent. Companies that would like to hedge the price risk of buying or selling a specific commodity often find that the benchmark price in a futures exchange tracks imperfectly. In these instances, the problem is usually resolved with a futures contract tied to the benchmark and an OTC contract that covers basis risk. Put another way, the demand for hedging in a futures market often depends on a reliable OTC to cover basis risk.

Exchanges ensure price transparency of key commodities, but cannot serve all the needs of the market. The volume of trade in commodities with unusual qualities or in isolated geographic markets is too small to be economic in exchanges, where there are strong economies of scale arising from large fixed costs. At the same time, however, the immediate price transparency of exchange trading guides OTC trading, providing reliable benchmark prices which are essential for calculating basis differentials. This feeds back and strengthens futures exchange trading. In other words, basis swaps made in the OTC directly link the two types of activity, ensuring greater trade volume and lower transaction costs for both.

A mature commodity market with well-established futures contracts complements OTC trade, providing buyers and sellers a transparent portrait of expected prices in future periods. These prices determine the fair market value of a long-term contract that might have a single flat delivery price. In immature markets without futures, the relationship between current spot and longer-term contract prices certainly exists, but it is not transparent and often appears random and unexplainable. The consequence is inefficient investment decisions and sometimes a barrier to entry. In early stages of development, the gas industry relied on complex and rigid contracts in order to obtain funding. No one would build a pipeline (and no bank would finance it) unless there was assurance of a supplier's willingness to supply gas and a buyer's willingness to consume it. While these arrangements are often necessary to underwrite the construction of a capital-intensive infrastructure, they are not flexible and have difficulty in accommodating unexpected changes in the economic environment. Moreover, the large scale of the projects requires massive financing, which may exclude all but the largest companies.

Summary of the Benefits of Energy Futures Markets

The 2008 financial crisis gave derivatives a bad name, but all swans are not black, most are white. Here is a brief summary of what futures markets do for energy markets and why they are important. First, futures evolved from agricultural cash exchanges that had a mismatch between a short harvest season and twelve months of consumption. In this manner, they could reduce risk and allow farmers and millers orderly planning. Likewise, reliable forward prices provide crucial signals for oil producers and refiners, allowing efficient production and processing decisions. Second, separating physical and financial trading significantly increases the numbers and types of counterparties and ensures the credit worthiness of participants, reducing the risk of default. Third, futures exchanges provide enormous liquidity and greater competitiveness by bringing new players into the market. Fourth, they allow participants broad latitude as to when to buy and sell, thus reducing the risk and cost of managing inventory. Fifth, and most important, futures exchanges vastly improve market information by providing an open system for price discovery and transparency. Futures markets achieve these objectives by standardizing contracts, trading in an open exchange, encouraging financial rather than physical closings, and disseminating price information. The emergence of NYMEX and ICE futures trading had a profound impact on the oil industry, and as long as oil is broadly produced and consumed, they will play a key role.

Notes

1. The most notable of the electronic exchanges was Enron-on-line, but for some years there were at least a dozen competing alternatives, including Dynegy-on-line and the Intercontinental Exchange (ICE). Almost all of the fledgling trading platforms failed, with the notable exception of ICE. There are, however, a variety of electricity exchanges, such as NordPool in Europe or PJM in the U.S. These exchanges are usually integrated with grid management and serve multiple functions.
2. As a practical matter, the terms of art "swaps" and "CFDs" are often used interchangeably. However, a swap is a more general term than a CFD.

3. The effect of a CFD can be achieved by two actions in a futures market, buying and then selling at a later date an offsetting contract. In futures, the timing of the offsetting transaction is at the option of the buyer. In contrast, most CFDs specify in advance the forward dates in which the contract price will be offset against the relevant spot price. For example, a CFD buyer may agree to pay (or be paid) the difference between the contract's fixed price and the floating price one month hence (there will be a specific date and source for the floating price). The price difference is then multiplied by the contract volume to determine who pays who and how much.
4. The oil market has seldom used flat prices in long-term contracts. Before the development of futures markets, term contracts usually tied contract prices to posted prices (described in chapter 3) which changed with market conditions. OPEC adopted a similar scheme when it took control of the international market in 1973. In contrast, long-term natural gas contracts in North America normal set flat prices until the mid 1980s. This was possible because gas was initially a waste product, sold into a regulated retail market. European and Asian natural gas contracts typically index prices to a market basket of oil products.
5. Some critics have complained that the high futures volume is excessive. There are, however, many different ways to interpret these numbers. Crude oil futures contracts extend for 6½ years into the future although most trading is in the first few months. Professor Ronald Ripple points out that comparing financial contracts that rapidly turn over to physical flows is largely an apples-to-oranges comparison. Future contracts are traded daily, but obligated deliveries are over a full month. Dr. Jeff Currie of Goldman Sachs notes that if producers are seeking to hedge reserves, rather than daily production, the annual volume of futures trading is only a small fraction of the nearly trillion barrels of reserves in the ground.
6. The brokerage cost of trading oil futures is remarkably cheap, 2¢ per barrel or less.
7. Oil production actually exceeded oil consumption by 2 million barrels per day in 1979. At the time, there was huge fleet of surplus oil tankers, induced by a sharp drop in oil imports, following the 1973-74 oil shocks. These ships were the primary means of oil storage. Since then, the number of surplus tankers has declined and there is limited spare capacity.
8. There is an active spot and futures natural gas market in the U.K., but unfortunately, this type of market organization does not extend to the main continent.

5

THE NEW INDUSTRY STRUCTURE

The Big Bang Theory of the Universe

Most physicists subscribe to the *big bang theory* of the universe's origin, resulting in an ongoing expansion. Opinions on future events are more diverse; will the universe continue to expand or will it expand, contract, and then expand again? The latter theory is not popular, but it has a tidy logic. More importantly for the analysis here, that logic fits a cyclical industry. Largely, it is the experience of the oil industry.

Figure 5–1 depicts what Joseph Schumpater (an economist writing in the time of the Great Depression) might have identified as "creative destruction," that is, capitalism's ability to recast its principal activities as circumstances and opportunities change. Figure 5-1 encapsulates a diverse set of events during the last four decades. In so doing, it invariably omits details and anomalies, but it is useful vision for gaining perspective and helps to explain the present petroleum industry structure.

The background of the chart is significant. The shaded area of figure 5–1 represents the total value of crude oil produced during each year, adjusted for inflation (2007 dollars). From 1965 through 1972 surplus Middle East crude oil and low prices kept the value low; during the period total value only rose from $138 to $242 billion. At the time, a few publicly owned corporations dominated oil trade. The corporations sprang from just three countries, the U.S., UK, and the Netherlands. They had familiar brand names: Exxon, Mobil, Shell, BP, Chevron, Gulf, and Texaco; nicknamed, the Seven Sisters by the Italian, Enrico Mattei, anxious to join the club. During this period, OPEC member governments chaffed at the Seven Sisters' market control and independent companies fought for access.

Figure 5–1 is also a timeline of changes in industrial structure. In general, the bottom part of the chart identifies the number and types of significant oil companies *created*. In most cases, host governments sponsored the new entities, reflecting concern about control over resources or markets. The first wave of new companies was a consequence of the OPEC revolution. The second wave came along several years later with the breakup of the Soviet Union and China's shift away from state-owned enterprises.

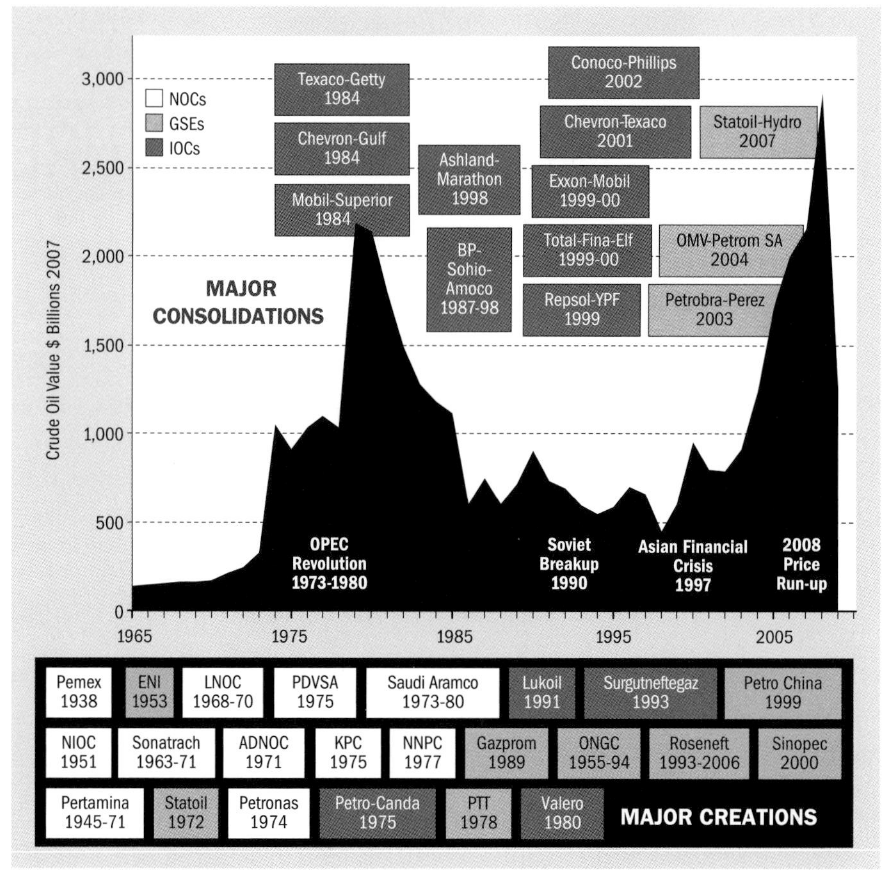

Fig. 5–1. Shifting industry relationships and structure 1965–2009

OPEC members' began to assert control in the early 1970s following the example set by the Mexican government, which had created Pemex as a national oil company in 1938. As a general rule the OPEC governments either created companies from the components of their petroleum ministries or they nationalized (or purchased) the assets of companies operating in their country. In all cases, results were similar—a company solely owned and operated by the government. In 1980, the Saudi government completed its purchase of ARAMCO signaling the final break between the interests of the major oil companies and OPEC's host governments.

The enthusiasm for direct government control of important energy supplies was not limited to OPEC. The 1973 Arab oil embargo was a shock to the oil consuming nations. Governments that could do something about securing supply did so—The UK created the British National Oil Company (BNOC,) Japan created the Japanese National Oil Company (JNOC,) Canada created Petro-Canada, and Norway founded Statoil. Later on, Prime Minister Thatcher dismantled BNOC. JNOC limped along until 2004, when it was finally euthanized. Petro-Canada, received subsidies and special resource access for years; it was finally privatized in 2004. Statoil survived handily with a mix of government and private ownership, in part, because it was a pioneer in deepwater exploration and could draw on Norwegian marine expertise.

The top of figure 5–1 depicts the other extreme of industrial re-organization, *consolidation*, primarily through mergers and acquisitions. In response to OPEC's resource control and the changing fortunes of the industry, many publicly held oil corporations merged. In 1979, the value of produced crude oil topped out at $2.2 trillion in current dollars (it would not reach this level again for 28 years). Over the next seven years, there was a substantial contraction in annual crude oil value and widespread consolidation. In 1984, Chevron acquired Gulf and a few months later Texaco acquired Getty. Exxon acquired Superior, while the remaining majors swallowed up other independents. At the time, the companies were accused of drilling for oil on Wall Street.

Following the collapse of oil prices in 1986, the total value of crude oil production fell to $601 billion. Low oil prices caused BP to consolidate its holdings of Sohio, thus gaining full control of half of Alaska's oil. Standard Oil of Ohio had been Rockefeller's original company, and a key part of the Standard Oil Trust. For many, BP's removal of the Standard Oil sign high above Public Square in Cleveland represented the culmination of century-long competition, an ending that surely made Standard's founder turn over in his grave.

On November 9, 1989, East Germans breached the Berlin Wall and within a few months, the Soviet Union imploded. Also in 1989, the Soviet Ministry of the Gas Industry reorganized into Gazprom. Three years later the Russian Federation chartered Gazprom as a joint stock company and by 1995 eager stockholders flooded its annual meeting. Russia had done an extraordinary about-face. Within a few years, its infamous oligarchs carved up the state's vast resources. The process created a number of private companies, which consolidated into two giants, Lukos and Surgutneftegaz. The Russian Federation also created a national oil company, Roseneft, which struggled until 2000. Since then Roseneft was partially privatized, placed under new leadership, and is now a financial success.

The breakup of the Soviet Union freed substantial quantities of oil and gas from inefficient uses. The flood of new supply in combination with the Asian financial crisis of 1997–1998 sent the annual value of crude oil down to $448 billion in 1998, its lowest level since the OPEC revolution. The reduced cash flow provoked another wave of consolidation among the remaining publicly owned corporations.

In 1997, two independents, Ashland and Marathon, agreed to merge. A short time later BP acquired Amoco (another Standard Oil company). These mergers and the rumblings of more to come touched off a major consolidation in the

European oil market. The French oil company, Total, first moved to acquire the Belgium oil company Petrofina, fighting off ENI and Elf Aquitaine who were competing suitors. The newly formed company Total Fina then launched a hostile bid for its beaten competitor Elf Aquitaine. The bid was eventually successful and the new company Total Fina Elf S.A. became one of the largest oil companies in the world. In 1998, Repsol, the Spanish oil company, acquired 14% of YPF, Argentina's national oil company. A year later Repsol acquired the remaining assets and the new company became Repsol YPF S. A.

The merger wave continued when in November 1999, Exxon and Mobil announced a merger, recombining parts of the old Standard Oil monopoly. Two years later Chevron and Texaco merged, reducing the original Seven Sisters to four, but now with a new co-equal in Total Final Elf. There were other consolidations too: BP swallowed ARCO's refineries and distribution system. Chevron acquired Unocal. Shell acquired Pennzoil in 2002. Previously Shell and Texaco had been in partnership with Saudi Arabia for a joint refining and gasoline-marketing venture, known as Motiva. The Chevron-Texaco merger spun off the Motiva assets to Shell, which ironically had not been part of the original ARAMCO consortium. In addition, a number of former independents banded together. Most importantly, Conoco, Phillips, the refiner and marketer Tosco, and the Alaska resources of ARCO combined to form ConocoPhillips, which is now the sixth largest oil company in the world.

In the late 1990s, China undertook a major restructuring of its energy industries. Previously, the government had divided the oil industry between China National Petroleum Corporation (CNPC) and the China Petroleum and Chemical Corporation (Sinopec). CNPC was responsible for domestic exploration and production, while Sinopec handled petroleum product refining and distribution. In the revised structure, China sought to create two vertically integrated companies each with a regional base. China never fully implemented the plan; instead, the government embarked on a multi-year process to transform state-owned enterprises into modern corporations. The idea was to strip bloated state operations of the best assets and create efficient vertically integrated companies with a mix of public and government ownership. Three companies followed from this model: China National Offshore Oil Corporation (CNOOC,) Sinopec itself, and Petro China (primarily a spinoff of CNPC).

The consolidation of IOCs and the entry of powerful new competitors from China reverberated throughout the industry. In 2003, Petrobra expanded its holdings in South America by acquiring the Argentine company Perez Companc Energia. In 2004, OMV acquired Petrom SA which included most of Romania's oil producing assets. Previously OMV had expanded its refinery and distribution system from Austria to neighboring countries following the breakup of the Soviet Union. The last major merger of the era was in Norway when Statoil and Norsk-Hydro agree to merge. The Norwegian government owned a portion of both companies and both had extensive operations in the North Sea.

Despite the reorganization, the global market share of the four remaining Sisters has continued to decline. A variety of expanded activities and new organizations fill the vacuum. Service companies have grown rapidly to provide development support for oil and gas activities all over the globe. With rising oil prices, the national oil companies of the oil exporting nations have prospered, and some aim for international prominence by developing projects outside their protected markets. Newly minted companies like Petro China became household names in a few years and have special access to what likely will become the largest petroleum market on the globe. It's a big stretch to pronounce the demise of the remaining Sisters: ExxonMobil, ChevronTexaco, BP, and Royal Dutch Shell, but there is less swagger in their step. As the power of the major international oil companies has eroded, it translates to rising dependence on OPEC's national oil companies for crude oil supplies.

Defining the Industry Structure

With the cycle of consolidation and expansion, new terms of art have evolved to describe the industry. Instead of "majors." the large multinational oil companies are commonly referred to as International Oil Companies or *IOCs*. The IOCs are listed corporations with 100% of their stock owned by the public and actively traded. The public knows a great deal about these companies; they publish annual reports and abide by SEC and other regulatory rules. In contrast, knowledge about the companies charted and owned by national governments is sketchy at best. For want of a more romantic term, a mundane label applies—National Oil Companies or *NOCs*. The ordinariness of the title, however, cloaks their operations. Sometimes they are little more than engines of national revenue—taxmen by another name. Most of the companies hide profitability, operating revenue, numbers of employees, and other key variables from public view, and, in a few cases, the information is a state secret.

In addition to IOCs and NOCs, a third class of oil companies became increasingly popular in the last decade—companies that are part public and part government owned. U.S. officials define Fannie Mae and Freddie Mac mortgage companies, as Government Sponsored Enterprises or GSEs. This is an apt description for companies like ENI, from Italy, OMV from Austria, and Petrobras from Brazil. The publicly owned portion trades on stock exchanges and most GSEs provide key data in annual reports. In some instances, GSEs have shed all previous government ownership. (Prominent examples are BP, Total, Repsol, and Petro-Canada.) Figure 5–1 classifies companies based on their present status. NOCs are in white, IOCs are in dark gray, and GSEs (as a blend of government and public ownership) are in light gray.

Typically, IOCs operate in oil-importing countries where they manage large refinery and distribution systems. Almost all have a long and varied corporate history. They also develop large-scale oil and gas production projects in countries where the local NOC lacks the technology and financial resources to undertake the development. Most NOCs stay close to home, developing local resources with the aid of international service companies. Only Kuwait has been successful in refining and marketing products in Europe, Asia, or the Americas under the brand Q-8. The various OPEC governments created NOCs in the 1970s when they successfully broke away from the major oil companies' influence.

GSEs, as the acronym suggests, are also the stepchildren of government policy, but their structure provides far greater flexibility. These companies actively seek investment funds and have a structure that allows them to access major capital markets. In the case of Russia, there were extensive resources but the fields were in decline and major upgrades in facilities were essential. Russian IOCs, such as Lukoil, were able to tap western capital markets and expertise. Roseneft, as an NOC, fell behind such private development. Once recast as a GSE, however, Roseneft had adequate corporate governance, an incentive to become efficient, and an opportunity to access capital markets. It also had an obvious political and resource advantage because majority ownership remained with the Russian government.

Although GSEs structure themselves like modern corporations, their motivations are more complex. The public expects them to be profitable and, at the same time, serve broader political goals. Petro China is good example. China National Petroleum Corporation owns the bulk of Petro China, and the government owns all of CNPC. Petro China serves a number of purposes—it is a bridge between China's state-owned enterprises and an emerging class of publicly owned corporations. Both Petro China and CNPC manage resources in China and invest abroad to secure the country's oil imports. Petro China has a broader agenda than just profitability. The company serves the national interest of China, where government policy makers seek to secure reliable oil imports.

Exxon Mobil remains the largest private corporation in the world, but for a time in the fall of 2007, Petro China, had a larger market capitalization (measured on the small proportion of the company actually traded). The *Financial Times* (FT) listed 43 oil- and gas-producing companies among its top 500. This group of companies had a total market capitalization of $3.9 trillion and an annual turnover, or sales, of $3.2 trillion in 2007. The four remaining sisters—Exxon Mobil, Royal Dutch Shell, BP, and Chevron—represented only 27% of the FT oil and gas group's market capitalization, and 39% of total sales.

Table 5–1 lists the 16 largest investor owned oil and gas companies, ranked by 2007 sales. The top companies today were at the top in the 1970s, but only four survive. Today six companies can be classified as traditional majors, with Total and ConocoPhillps joining the rarified atmosphere reserved for former

Sisters. The remaining companies are smaller and typically more specialized by sector or geographic area.

Table 5–1. IOC rankings by 2007 sales

FT Sector Rank	FT Global Rank	Company	Country	Market Value 12/31/07	Sales 2007 Billion USD	Net Income Billion USD
1	1	Exxon Mobil	US	452.5	390.3	40.6
4	9	Royal Dutch Shell	UK	220.1	355.8	31.3
6	16	BP	UK	191.8	284.4	20.8
7	18	Total	France	178.6	216.3	20.8
8	19	Chevron	US	177.3	214.1	18.7
11	45	ConocoPhillips	US	119.0	187.4	11.9
38	349	Valero	US	26.3	95.3	5.2
23	193	Repsol-YPF	Spain	42.3	88.4	5.0
15	89	Lukoil	Russia	72.7	67.7	7.5
29	274	Marathon Oil	US	32.3	64.6	4.0
32	300	CEPSA	Spain	29.7	33.6	1.2
36	321	Hess	US	28.3	31.6	1.8
20	240	Imperial Oil	Canada	47.4	24.7	3.1
18	128	EnCana	Canada	57.2	22.7	4.0
42	441	Petro Canada	Canada	21.1	21.4	2.7
25	216	Surgutneftegas	Russia	39.4	21.3	3.3

Source: *Financial Times*.

The NOCs listed in table 5–2 are mainly from oil exporting countries. Precise sales figures are mostly unknown. For oil exporting countries, the figures in the table represent the country's production times the average annual oil price. Some caution should be used in interpreting the table. For example, NNPC has a limited role in Nigerian oil production, while Saudi Aramco has complete control of Saudi oil production. Nonetheless, the table provides an overview of the more active NOCs.

As noted, the GSEs listed in table 5–3 were designed to meet a variety of national and corporate goals. At one end of the spectrum, Petrobras, Statoil-Hydro, Gazprom, and Roseneft, were founded on the intention to manage and control the sponsoring country's oil and gas development. Nonetheless, some of these companies like Statoil-Hydro pursue development in other oil-producing regions, particular where they have a leading technology, such as deepwater development. At the other end of the spectrum companies like OMV were designed to manage the country's refinery and distribution capacity and procure imported oil.

Table 5-2. NOC rankings by estimated 2007 sales

Company	Country	Estimated Sales Value 2007 Billion USD	Net Income Billion USD (when reported)
Saudi Aramco	Saudi Arabia	267.2	
CNPC	China	137.1	18.4
NIOC et al	Iran	121.4	
Pemex	Mexico	104.5	-1.5
Adnoc	UAE	77.8	1
PDVSA	Venezuela	73.0	6.3
Sonatrach	Algeria	67.5	9.3
NNPC	Nigeria	67.2	0.1
KPC	Kuwait	66.7	5.9
LNOC	Libya	51.5	
Sonangol	Angola	45.5	1
KMG	Kazakhastan	39.0	3.3
QP	Qatar	34.7	0.3
Pertamina	Indonesia	31.2	2.7
Petronas	Malaysia	27.3	9.4
Socar	Azerbaijan	22.3	

Source: *BP Statistical Review of World Energy*, July 2008; company Web pages and annual reports, where available; otherwise estimates from country volumes and market prices.

Table 5-3. GSE rankings by 2007 sales

FT Sector Rank	FT Global Rank	Company	Country	Government Ownership	Market Value 2007 Billion USD	Sales 2007 Billion USD	Net Income Billion USD
10	37	Sinopec	China	76%	135.3	148.9	7.7
9	36	ENI	Italy	30%	137.1	137.8	15.8
2	2	PetroChina	China	86%	424.0	119.0	20.8
13	66	Statoil-Hydro	Norway	63%	95.8	102.5	8.7
5	12	Petrobras	Brazil	32%	208.4	98.5	12.4
3	4	Gazprom	Russia	51%	299.8	91.6	26.1
35	320	PTT	Thailand	52%	28.3	47.1	3.1
12	65	Rosneft	Russia	89%	95.9	33.1	3.5
43	478	OMV	Austria	32%	19.9	30.0	2.2
19	148	ONGC	India	84%	52.3	20.5	4.4

Source: *Financial Times*, CEE, Web sites.

An Aside on Transparency and Public Responsibility

One poignant irony emerged following the breakup of the Soviet Union. It turned out that the environmental problems in the member countries were astonishingly deep and protracted, representing years of neglect. Given the central government's ability to direct resources exclusively for the "public good," it is surprising that they would neglect the environment. It turns out that the capitalist system of corporate transparency, public pressure, and regulation is much better suited to curb excesses than the Soviet system of central planning.

A parallel story emerges regarding oil development in the Sudan. Chevron originally discovered the oil. It is of good quality, but a long distance from tidewater and the export market. Chevron also ran into a series of problems resulting from the suppression of Christian and Animist minorities in Southern Sudan by the Moslem government in Northern Sudan. The resulting disruptions and bad publicity caused Chevron to sell its oil interests to BP, which quickly ran into the same problems. BP sold out to three former employees, but they did not have the financial depth to provide development. The wildcat explorers formed a joint venture with an independent Canadian producer, Talisman. In turn, Talisman brought in CNPC and Petronas, national oil companies of China and Malaysia.

Sudanese oil development turned out to be complicated. The unsavory reputation of the Sudanese government made it impossible to finance the necessary pipeline and infrastructure development through normal channels. Financing proceeded by barter. For example, China supplied the steel pipes. The timing of the venture was lucky; the oil came on stream just as prices began to rise. Unfortunately, the repercussions were less providential. Critics claim that as the first cargo of oil left Port Said, it passed an inbound freighter, containing tanks and other military equipment. The factional war blossomed into a deadly dispute, linked, in part, to the control of oil.

The foreign partners in the Sudanese oil venture, with the exception of Talisman, were NOCs. Chevron and BP had retreated from the bad publicity and ultimately Talisman did too. Despite the venture's profitability, stockholders brought considerable pressure on the Board to withdraw. Within two years, Talisman sold its interests to another company; this time to ONGC, primarily owned by the government of India.

Oil production and the revenue from it sustain the Sudanese government's ability to carry on its war policies independent of UN sanctions and general condemnation. For the NOCs active in Sudan, their governments' concern over supply security trumps international criticism. Profit-driven IOCs abandoned the project, driven away by public pressure and some sense of moral responsibility. The benefit from Sudanese oil production was obviously less than the destructive impact on their brand name. The irony here is even deeper than the Soviet mishandling of the environment. At least in the case of Sudan, for-profit capitalist companies turn out to have more sensitivity to social and moral issues than do NOCs.

The Rise of the Service Companies

The shift in resource control from vertically integrated multinational corporations to the governments of oil exporting countries has been the primary force behind industry restructuring. One aspect of the shift has been a locking out of companies traditionally responsible for the development and management of the world's largest oil fields. NOC's have taken over, but often they lack the expertise to manage and develop and consequently turn to a different type of company.

Table 5–4 lists the five largest energy service companies tabulated by the *Financial Times*. These companies grew rapidly through the 2005 to 2008 oil boom. In a major turn of events, Schlumberger's 2007 sales revenue *was greater than all but two IOCs*. As the IOCs consolidated and shrank the service companies grew. Figure 5–2 illustrates the impact in terms of number of employees.

Table 5–4. Service companies rankings by 2007 sales

FT Sector Rank	FT Global Rank	Company	Country	Market value 12/31	Sales 2007 Billion USD	Net Income Billion USD
1	57	Schlumberger	US	104.2	232.8	51.8
3	248	Haliburton	US	34.6	152.6	35.0
5	436	Baker Hughes	US	21.2	104.3	15.1
4	371	Weatherford Int.	US	24.6	78.3	10.7
2	190	Transocean	US	43.0	63.8	31.3

Source: *Financial Times*.

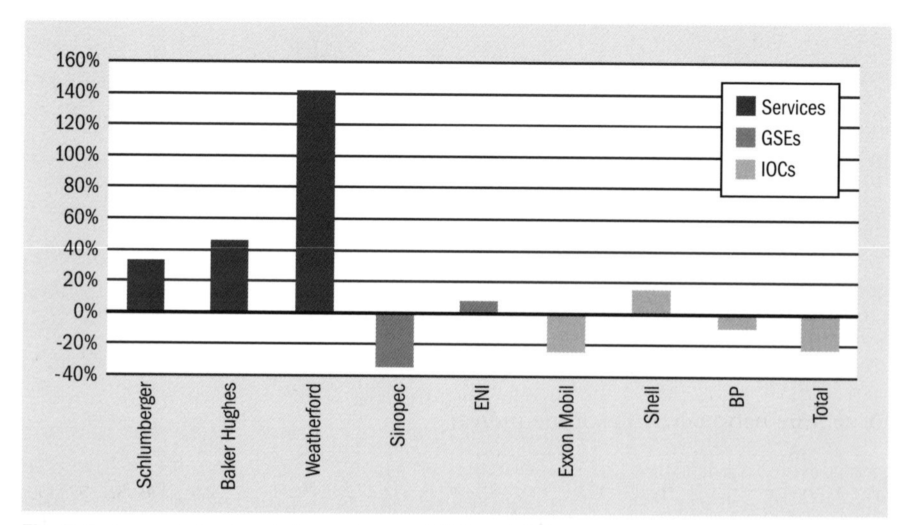

Fig. 5–2. Percentage change in employees 2000 to 2007

Royal Dutch Shell was the only IOC to increase the number of employees between 2000 and 2007, an increase of 16%. (Ironically, Shell was the only large IOC uninvolved in the merger mania.) In contrast, the service companies grew from 33 to 142%.

Conflicting Political Motivations, Incentives, and Opportunities

The international oil industry is a complex and dynamic combination of public and private ownership. Recently the tide has shifted toward far greater government participation or outright control of the companies that explore for and develop crude oil resources around the globe. Put another way, the NOCs and GSEs listed in previous tables are now the major players for future oil supplies. The Baker Institute at Rice University recently studied the industry, pointing out that Western international oil companies "now control less than 10% of the world's oil and gas resource base." (Baker Institute 2007, p. 1). Moreover, as a general observation, the reserves held by Western companies are of lower quality and will have a higher cost of development. The most promising oil and gas reserves and greatest potential for new supplies remain in the Middle East, which is unlikely to be opened up to IOCs in the way it once was.

Although the extent of government involvement is greater than before, the organizational paradigm is not new. Eight of the IOCs listed in table 5–1 have had partial or full government ownership at one time in their history. The histories of BP and Petro-Canada exemplify the motivations for government participation. In the case of BP, the motivation was clearly security of supply. Lord of the Admiralty, Winston Churchill, could not accept the risk that Royal Navy would be unable to obtain fuel oil during World War I. Ownership rights in the UK's largest oil supplier resolved that issue. In contrast, the Canadian government founded Petro-Canada due to concern about control over domestic resources. At the time, Canada's oil and gas reserves were declining (or not adequately increasing) due to large exports to the U.S. Many Canadians believed that they should husband their resources for the domestic market. Over time, the motivations for partial government ownership of both BP and Petro-Canada weakened as oil markets moderated and the number and types of suppliers expanded. Canada rediscovered that it was good business to sell oil and gas to energy-starved Americans. Moreover, subsidies to a state-sponsored energy company were a drain on the budget. Ultimately, the U.K., like the U.S., recognized that it had a variety of policy options to allocate fuels in times of emergency that did not require nationalization of the industry.

Because NOCs do not have to meet these standards set for listed companies, they may be highly inefficient. For example, the NOC of Japan, JNOC, was a dumping ground for mid-level bureaucrats and required subsidies throughout its existence. At its peak, CNPC had more than 2 million employees; the company did everything: it not only produced oil, it ran everything for its employees, from laundromats to movie theatres. The new Chinese energy corporations are far more efficient and profitable. The measure of profitability is, however, frequently suspect because many of these companies receive hidden subsidies. The subsidies most often take the form of preferential access to either resources or markets.

It is, however, worth pointing out that preferential access to resources has frequently been instrumental in the initial success of a number of previous GSEs that are now IOCs including ENI, Total, Respol, Lukoil, Petro-Canada, and Surgutneftegaz. Thus, the pendulum could easily swing back to more extensive private control and less government involvement. Many of the GSEs, and even some NOCs, could transition to IOCs in the coming decades. In any case, all the current GSEs receive some kind of special advantage or access that purely private companies have been denied. Thus, they seem firmly lodged in their positions.

There are times when political motivations may be in conflict with a firm's profitability. Is the mission of a GSE to maximize profits or to secure energy supplies for its sponsoring government? For example, Petro-Canada was required to make substantial investments in oil sands irrespective of oil prices or profitability. In this instance, the broader interests of the Canadian government to see the resource identified and developed overrode the company's for-profit motive.

The choices governments make in organizing their energy industry often come down to the opportunities presented. The popularity of the NOC structure in oil exporting countries is not accidental; it usually reflects a low-cost and abundant resource base. The NOCs don't need to fret over difficult and risky investment choices and they often don't necessarily need access to capital markets. Although they often face a dilemma—how to divert the cash flow from government spending to energy projects that would provide a high rate of return.

Recent trends in the development of GSE's have been quite different from earlier experiences, such as BP and Petro-Canada. Russia is the prime example of a resource-endowed country seeking to assert control over the pace of development and the rights to the associated revenue. The Russian experience is, of course, unique—shifting from complete government control to privatization and then back to a mix of government and private ownership and control. Interestingly, the period of privatization was highly productive for resource development. The government gave the newly formed Russian companies access and the incentive to apply the latest technology to aging oil and gas fields. The result was a substantial turnaround in oil production. Recently, the benefits of modernization have worn off. In addition, uncertainties over government taxes and other constraints have led to falling or stagnate production.

Likewise, the public-private mix of ownership of Chinese energy companies reflects broader issues of national policy: in this case, both issues of national security and the transitory path to a modern economy. The government considers energy vital to Chinese economic development and although the country has abundant resources of coal, it imports both oil and natural gas. One of the important functions of Chinese energy companies is to explore and invest in energy resources in a variety of locations around the globe in order to diversify imports and, on occasion, secure an equity advantage. Moreover, the GSE framework is natural for country transitioning to a market-based economy. The Chinese believe that the Russian experiment to dismantle the Soviet economy machinery virtually overnight created unnecessary hardship and extensive political unrest. They did not want to repeat the Russian mistake.

Vertical Integration

A snapshot of the oil industry today could not be more different from the tightly controlled market structure before 1973. Table 5–5 lays out crude oil production and refining capacity of the larger IOCs, GSEs, and NOCs, allowing measurement of vertical integration in the key sectors. Two of the six majors, BP and Chevron, coordinate resources in the traditional manner. That is, they have a balance between crude oil production and refining capacity. Three of the majors, ExxonMobil, Royal Dutch Shell, and ConocoPhillips, are weighted towards refining; that is they depend on a large volume purchase of crude oil to keep their refineries fully utilized. In the case of ExxonMobil, this may be mildly misleading because the company remains somewhat tied to Saudi Aramco, which it once owned along with Chevron and Texaco. In the case of all the majors, they are active natural gas producers and, as previously noted, now count gas reserves in crude oil equivalents. Thus, they are maintaining a reasonable balance between production and refining.

The remaining large IOCs listed in table 5–5 vary in their organization. Hess and Petro-Canada retain a vertically integrated structure, while Surgutneftegas is primarily a producer. Valero is a refiner and gasoline marketer, with around 3 million barrels per day of refining capacity. The Coastal Company spun off Valero as a separate entity in 1980. (Coastal was likely motivated to increase stock value by divesting unprofitable refineries.) Valero was able to acquire refineries because as the IOCs consolidated, they spun off assets that were duplicative or created antitrust concerns for regulatory agencies. Valero snapped them up.

Overall, the major IOCs listed in table 5–5 have crude oil production averaging about two-thirds of refinery capacity. In trading terms they are "net-short," depending on purchased crude oil to round out their refinery runs. The development of efficient markets for crude oil and petroleum products allow greater specialization than in previous eras. This is explained further in chapter 7.

Table 5–5. Degree of vertical integration

Company	Type	Country	2007 Oil Production kbpd	Refinery Capacity or Throughput kbpd	Degree of Integration
Valero	IOC	U.S.	–	2,875	0.0%
Respol-YPF	IOC	Spain	176	1,233	14.0%
Marathon Oil	IOC	U.S.	197	1,016	19.0%
EnCana	IOC	Canada	134	452	30.0%
ConocoPhillips	IOC	U.S.	854	2,560	33.0%
Exxon Mobil	IOC	U.S.	2,616	5,571	47.0%
Royal Dutch Shell	IOC	UK	1,818	3,779	48.0%
Imperial Oil	IOC	Canada	275	442	62.0%
Total	IOC	France	1,509	2,413	63.0%
Chevron	IOC	U.S.	1,783	1,833	97.0%
Petro Canada	IOC	Canada	297	297	100.0%
BP	IOC	UK	2,414	2,127	113.0%
Hess	IOC	U.S.	275	227	121.0%
Lukoil	IOC	Russia	1,941	1,135	171.0%
Sinopec	GSE	China	799	3,026	26.0%
OMV	GSE	Austria	164	528	31.0%
Petrobras	GSE	Brazil	1,920	2,227	86.0%
PetroChina	GSE	China	2,276	2,256	101.0%
ENI	GSE	Italy	1,020	743	137.0%
Roseneft	GSE	Russia	2,027	1,137	178.0%
Statoil-Hydro	GSE	Norway	1,070	325	329.0%
PDVSA	NOC	Venezuela	2,613	3,105	84.1%
Pertamina	NOC	Indonesia	969	992	97.7%
Petronas	NOC	Malaysia	755	545	138.6%
Socar	NOC	Azerbaijan	868	442	196.4%
NIOC	NOC	Iran	4,401	1,857	237.0%
Pemex	NOC	Mexico	3,477	1,463	237.7%
KPC	NOC	Kuwait	2,626	880	298.4%
QP	NOC	Qatar	1,140	350	325.7%
Sonatrach	NOC	Algeria	2,000	450	444.5%
LNOC	NOC	Libya	1,848	378	488.8%
Saudi Aramco	NOC	Saudi Arabia1	10,413	2,020	515.5%
NNPC	NOC	Nigeria	2,356	439	536.9%
Adnoc	NOC	UAE	2,915	500	583.0%
KMG	NOC	Kazakhastan	1,490	193	772.2%
CNPC	NOC	China	2,705	211	1282.0%
IOC Average					65.7%
GSE Average					127.1%
NOC Average					415.9%

Source: *Financial Times*, CEE, Web sites.

In contrast, the GSEs in table 5–5 are closer to the vertically integrated model, with crude oil production 122% of refinery capacity. There is, however, substantial variation with Sinopec's oil production only 26% of its refinery capacity and Roseneft's at 178%. It turns out that when governments invest directly in the oil sector, the primary factor influencing the degree of vertical integration is the at-hand opportunity.[1] Countries with a large resource base have companies that maintain a higher proportion of crude-oil production relative to refinery capacity. Countries with a limited resource base are far less likely to have a vertically integrated industry. Generally, attempts to secure energy supplies from external sources have not been successful, as Japan's experience with JNOC demonstrated.

The NOCs in table 5–5 have a crude-oil production capacity 416% of their refinery capacity. That should not be a surprise; the primary function of these companies is to manage oil production and export.

Notes

1. ENI, as the senior GSE in table 5–3, is an exception with multiple worldwide oil production investments.

6

ENERGY SECURITY

"A horse! A horse! My kingdom for a horse!" So laments the king in Shakespeare's *Richard III*. The sudden and urgent need for mobility runs the gamut from picking up the kids at daycare to a bank job get-away. In moments of stress, motorists feel highly vulnerable to any erratic behavior of energy suppliers. Similar anxieties do not surface for commodities considered less of a necessity. After all, in a commodity market, anyone can purchase the good, if they are willing to pay the price. In spite of everything, the anxiety about excessive reliance on foreign oil imports has never been higher. It is likely that many believe that in an emergency, markets will not be allowed to work, as has often been the experience. To add to the confusion, energy security is composed of a patchwork of issues. Advocates with motives that range from xenophobia to old-fashioned pandering argue to their narrow advantage and confuse the issue. The concerns most often expressed under the claim of energy security include: national defense, sovereign rights, and political extortion.

National Security

Originally, generals and admirals thought of energy security as a straightforward strategic objective aimed at ensuring fuel for national defense. The linkage was most obvious during World War II, when fuel shortages (and the fear of shortage) constrained both Japanese and German operations. During the war, fuel shortages also plagued important civilian activities. This problem surfaced again on a global level during the 1973 to 1974 oil crisis, giving rise to the International Energy Agency's oil allocation and strategic storage programs. In retrospect, the shortages experienced in the U.S. in the 1970s, or China and India in 2008, were primarily the product of shortsighted attempts to control prices. No such problems emerged in countries that allowed prices to balance demand and supply.

On the eve of World War I, Winston Churchill modernized the British fleet, converting its boilers from coal to oil. The conversion, however, created a new problem—securing oil supplies, because the UK had no domestic oil production. The problem was resolved when the British government acquired a 51% stake in the Anglo-Persian (British Petroleum) oil company (Yergin 1991 pp. 153–164). In pursuing the acquisition, oil prices and financial issues were a consideration for the Royal Navy, but the driving motive was security of supply—the physical availability of oil.

With its abundant domestic production, the U.S. had fewer reservations about switching its Navy from coal to oil and did so before the British; nonetheless, security of oil supply was a major consideration. Reserves from the Elk Hills and Buena Vista oil fields in California and the Teapot Dome in Wyoming were set aside in 1912. It is interesting to note that the reserves led to the famous "Teapot Dome Scandal" in 1922, a common outcome when private and public interests are mixed. As a consequence of the scandal, most of the reserves remained locked in place until the oil market shocks of the 1970s. The Clinton Administration finally sold the Elk Hills field to Occidental Petroleum in 1998 (ironically, at the bottom of the oil price trough). The U.S. also divided the country into five Petroleum Administration Defense Districts (PADDs), and during World War II sponsored the development of a series of domestic pipelines. The pipeline system aimed at minimizing tanker exposure from enemy submarines, but it also enhanced continental trade and became a keystone in the natural gas market's development.

The U.S. and the UK were not the only countries concerned about security of oil supplies, nor the only countries to do something about it. During World War II, the Allies had a significant advantage over the Axis powers arising from their control of vital sea lanes and the United States's substantial domestic supplies. Neither Germany nor Japan had significant oil deposits and both relied on sources that ultimately proved far beyond their control. Contemporary documents reveal that the need for secure oil supplies was a motivating factor in Japan's decision to bomb Pearl Harbor and the Nazis' decision to invade Russia. (Stern 2005, Becker 1981)

Even before the war started, the Germans had developed several synthetic coal-to-oil processes. Only one, however, (the "Fischer-Tropsch" process) was able to manufacture higher-octane gasoline. Germany began the war with a synthetic oil production capacity of 9 million barrels per year. It increased capacity to 36 million barrels per year by 1943 and met the majority of country's demand. Even then, however, their conversion process would not allow the production of 100-octane gasoline, which was required for the most advanced aircraft. Moreover, the facilities proved highly vulnerable to bombing in 1944, when the Allies destroyed 90% of the capacity. Germany never recovered (Becker 1981).

The first Gulf War marked the beginning of a new and disturbing threat associated with the supply of oil. In August 1990, Saddam Hussein invaded Kuwait. Political rationalizations aside, it was an outlandish grab of his neighbor's wealth and a throwback to meaner times. Kuwait produced 1.5 million barrels per day and had vast oil reserves: a jewel for the crown of ancient Babylon.

There is a subtle assumption that underlies the theory of "gains from trade." Namely, for both sides, the benefits of voluntary exchange must outweigh the gains from plunder of one by the other. This is indeed the case for most tradable goods, which require wit, risk taking, and experience to produce and market. The days of warlords storming a neighbor's storehouse of jewels and gold are mostly over. In the modern world, plundering undoes the means of production, and the plunderer gains nothing but an empty shell. This is not the case for many oil deposits, which often have vast economic and monopoly rents. At today's oil price levels, the lifting cost of producing most Middle East oil is frequently less than 5% of its value. Moreover, the technology of producing from an existing oil field is not complex and readily available. Think of these resources as vast storehouses of liquid gold, where control levies unimaginable wealth on the lucky owner irrespective of managerial skills or effort. In other words, in the case of oil, the potential benefits from plunder may easily outweigh the expected costs. Some believe that dependence on Middle East oil necessitates intervention and policing by the U.S. and its allies. If so, this is nothing less than a costly subsidy for imported oil.

When the oil industry was immature, supply deliveries were uncertain, transportation systems were unreliable and prices were opaque. Thus, there was a genuine concern about the physical availability of the commodity for national defense, industrial uses, and general consumption. As World War II revealed, however, a "secure" domestic infrastructure, such as the one developed by the Germans, may prove highly vulnerable to hostile attack, and recently in North America, to the destructive power of hurricanes. In any case, the early oil industry stands in sharp contrast to today's sophisticated commodity market. The national security issues related to oil supplies spring from the same specific asset problems that hobble competitive trade. Navy ships, tanks, logistical support, and so on all depend on the availability of particular type of fuel. Of course, there are substitutes for oil, but it takes time and considerable expense to make the adjustments.

Sovereign Rights

Although security of energy supplies is often associated with national defense, it can often arise from broader political or economic motivations. When South Africa pursued its policy of apartheid and shabby treatment of the black majority, it faced international condemnation and increasing isolation. Consequently, it developed a synthetic oil industry based on the Fischer-Tropsch process, and, thanks to its abundant low-cost coal, ultimately was able to produce fuels at costs not too much greater than prices charged by OPEC. South Africa achieved energy independence, but not without government subsidies and environmental cost.

In parallel fashion, decades of low oil prices and unfettered exploitation of crude oil reserves exasperated the leaders of many oil-exporting countries. The frustration led to the formation of OPEC. The motivation of OPEC's founders is often misunderstood. The idea was not to form a cartel and raise prices. Instead, the original goal was to reclaim control over each country's own resources. OPEC's members did not have mineral rights vested in private hands as understood in the developed world. Instead, each country's ruler granted concessions to explore and develop mineral resources with royalties and other benefits flowing to those governing. Thus, the formation of OPEC and its subsequent actions rest squarely in the political, rather than commercial, arena. This has had a significant impact on the oil and gas industry's development within these countries and, in turn, on the structure of western oil companies. Indeed, the majors' loss of OPEC oil is the pivotal event of the modern industry.

Political Extortion

With sovereign rights established it was an easy step for OPEC's leaders to view access to oil as political leverage, useful in meeting other objectives; extortion on a grand scale. The "oil weapon" emerged in 1973 when Arab OPEC (the Arab members of the Cartel) embargoed oil to the U.S. and the Netherlands. The provocation for the embargo was U.S. policy regarding Israel. It turned out that the embargo was not very effective—oil from other countries could quickly displace oil from the Arab producers. Oil is a fungible commodity, and once loaded on a tanker ships easily to most of the world's refineries. Regrettably, a production cut accompanied the embargo, which was how OPEC discovered its latent power over price. No country has repeated such a blatant act. Nonetheless, the importance of access to Middle East oil and particularly oil from Saudi Arabia has been a crucial element in U.S. policy since then, and the oil weapon remains a subtle threat.

Actions to cut off natural gas supplies can be far more effective because of the fixed delivery infrastructure. Europe has suffered a natural gas supply crisis twice in recent times, once in 2006 and again in 2009, centered on disputes between Russia and Ukraine. Russian supplies approximately one-fifth of Europe's gas with the majority of gas consumed in former Soviet Union countries. More than half of the gas flows through the Ukraine pipeline transit corridor and Ukraine itself depends on Russian gas. State-owned Gazprom (Russia's principal natural gas supplier) has been in dispute with Ukraine's distribution company, Naftogaz, over prices to be paid. Gazprom attempted to flow gas through Ukraine, and still cut off local consumption. Once the gas left Russia's border, however, they could control its distribution. According to Russia's leaders Ukraine was "stealing" gas meant to transit to other countries. Thus, Gazprom cut all gas at the Ukraine border, which left some countries like Bulgaria and Slovakia completely without gas.

Although Russia's leaders justified their action by claiming underpayment and theft, there were also broader political motives. Ukraine is seeking a closer relationship with the European community and NATO. But, at the same time Ukraine hosts the Russian Navy's Black Sea fleet at Sevastopol, which is a port that the Russians wants to keep. No matter who was at fault, Russia has undermined its own commercial interests. By curtailing supplies to countries not involved in the dispute, Gazprom now appears to be an untrustworthy supplier. Following the 2006 disruption, European countries began planning for alternative supplies, particularly LNG, which provides a number of supply sources and far greater flexibility.

The mix of political and economic motives that permeate international trade in energy products has a substantial impact on the industry's structure. Political risk is an important calculation in decisions about where to invest and whom to trust. This has advantaged some and disadvantaged others. Most importantly, security issues impede the efficient allocation of resources. Frequently, policy makers choose to develop high cost domestic resources, rather than rely on cheaper foreign sources from suppliers perceived to be undependable. Likewise, the managers of energy companies find it difficult to separate political considerations from their economic decisions.

7

ORGANIZING PRINCIPALS

Form Follows Function

"Form follows function" was the mantra of architects Louis Sullivan and his assistant Frank Lloyd Wright. The description was apt as building design broke away from classical styles and adapted new materials and construction techniques. The skyscraper was the consequence of a natural evolution driven by increasing urbanization, new technologies, and new building materials. Similarly, students of the petroleum business, such as Paul Frankel, argue that commodity's unique attributes drive the oil industry's form. In Frankel's view, organization, competition, investment decisions, product shape, and delivery systems all flow from the nature of the petroleum resource itself. Other economists are less persuaded, noting that the petroleum industry has reshaped itself many times in response to changes in regulation, property rights, or technology. Nonetheless, an examination of the petroleum industry's structure must begin with the idiosyncrasies of petroleum. What, if anything, makes the oil business, or any energy business, different?

Continuous flow

Petroleum industry executives describe the business as one of "continuous flow." Crude oil and natural gas flow from fields into pipelines that transport the commodity to processing centers and refineries. In a refinery, crude oil is converted to three primary products—gasoline, middle distillates, and heavy fuel oils. (A sophisticated refinery produces upwards of one hundred different products, ranging from coke to petrochemical feedstocks, but the main products remain gasoline and middle distillates). The refineries' products, in turn, flow to distribution centers, service stations, and so on. The oil never slows down. There are working storage tanks in the fields, at refineries and distribution centers, and at points of final usage. Typically, however, the amount of oil in storage will cover only a few days of demand. The primary role of working storage is to prevent logistical irregularities from creating spot shortages. There is one modest exception to the production and consumption cycle. During the spring, surplus gasoline is stored for the peak summer driving season and during the fall, heating oil is stored for the winter. The shoulder periods of refining activity are also the time for scheduled refinery maintenance.

Storage

The primary reasons why large amounts of oil are not stored are twofold. First, oil fields themselves are storage vessels. At any one time, companies hold a portfolio of undeveloped or underdeveloped resources, awaiting the right technological advances and economic environment. While it is not feasible to increase production beyond a field's design capacity in the short-term, field engineers can slow it down, or in industry jargon, "shut in" the oil flow. The most obvious examples are the cutbacks in OPEC oil production, discussed in chapter 3, that were aimed at preserving the cartel's prices. On occasion, oil wells have been shut in because the lifting cost of extracting and moving the oil is above its market value, but this is rare.[1]

When prices are low, there is little incentive for commercial companies to shut-in a field because, when reopened, production is constrained to the field's daily capacity, which is only a small proportion of reserves. In economic terms, this means that the shut-in barrel goes to the "back of the line." Its price must be discounted multiple years; reducing its present value significantly compared to its expected future price. Consequently, the current price must be very low, compared to the expected future price, in order to make production cuts worthwhile. This can be the case in a very weak market or when myopia sweeps the industry and producers develop unrealistic views of future oil value.

Ideally, there should be spare production capacity in order to meet unexpected fluctuations in demand or supply and to moderate price peaks and valleys. Such capacity is costly, however, and seldom seems worthwhile to individual producers. In the 1950s and 1960s, managing spare capacity was largely the task of the Texas Railroad Commission, which set pro-rationing limits on Texas and Louisiana oil producers. The U.S. occasionally used the surplus to offset oil shortages in the international market. For example, Texas and Louisiana supplied oil to U.S. allies during the 1957 Suez Crisis. Following the Iranian revolution OPEC members carried a large amount of spare capacity in an attempt prevent a complete price collapse. The spare capacity effectively constrained price spikes, with the exception of the period of the first Gulf War. When OPEC's spare capacity diminished to negligible levels, prices began to rise, peaking in 2008.

A second reason for skeleton oil inventory levels is the high cost of surface storage. These costs include the capital costs of the facilities—tanks, pumps, pipelines, and so on—and the foregone interest on the asset value of the oil. (Unlike shut-in production, the interest rate cost of carrying oil for the short-term is inconsequential; there are few effective short-term capacity constraints). The central problem with short-term storage is price risk. Of course, if prices are rising fast enough, the added value may more than cover the cost of stockpiling, but the industry cannot usually count on rising prices. In any case, it is not economic for firms to build seldom-used storage facilities.

In addition to the working inventories maintained by the industry, the U.S. government and fellow members of the International Energy Agency (IEA) agreed to maintain strategic oil reserves. Following the 1973 to 1974 Arab oil embargo, Congress authorized a strategic petroleum reserve (SPR). The oil is stored in salt domes, which is cheaper than tanks, but still expensive. These reserves are strategic, as the name implies. There is no intent to manage prices or meet other economic needs. Instead, the purpose of the reserves is to ensure the physical availability of oil in event of oil disruptions, due to revolutions, natural catastrophes, or other unplanned events. The immediacy and efficiency of today's oil market calls into question the purpose of the SPR because physical shortages would only be sporadic if prices balance demand and supply. Given that the biggest oil disruption risk is economic, rather than strategic, the costs may exceed benefits. Studies by Jerry Taylor and Peter Van Doren (2005) of the Cato Institute, suggest that the program is costly and seldom used.

Other examples of continuous flow

The oil industry is certainly not the only industry with continuous flow. Electricity generation and distribution on a centralized grid is the extreme example. The grid manager must balance demand and supply minute to minute. Natural gas also has continuous flow from the field to the consumers, but the gas grid is more flexible than the electric grid and real-time balance is not necessary. Natural gas grids, like power, can collapse if draw exceeds over input, but it takes hours or days. If the pressure drops too much, the grid must be shutdown and brought back step-by-step, as with an electric grid. Ironically, natural gas is cheaper to store than oil. This is because companies can refill depleted gas fields and use them as if they were underground storage tanks. However, in the case of both electricity and natural gas, a fixed infrastructure of wires or pipelines links producers and consumers directly. The delivery infrastructure is, without question a natural monopoly. Thus, as these industries have been deregulated, transportation systems have been set aside as regulated entities. Regulators foster competition by allowing producers, marketers, and distributers "open" or "third party" access to the regulated lines.

Traditional reasons for vertical integration

Paul Frankel reasoned that the oil industry's continuous flow and high storage costs led to vertical integration. The reasoning is straightforward. A producer must have an outlet for the crude oil since it is too costly to store or shut in. Likewise, a refiner seeks a secure source of crude oil, given the huge capital cost associated with refining. Moreover, there are spatial considerations. A specific transportation infrastructure links the two entities, which limits the number of competitors. Recall that Paul Frankel argued that oil, as a liquid, required specialized handling and such facilities had strong economies of scale. This in turn led to concentration in transportation and the anxiety felt by producers over secure outlets and refiners over secure supplies led to vertical integration. According to Frankel, this was the natural order that led to an industry dominated by a few large vertically integrated companies. Since then, the cost structure of the industry has changed, a commodity market has come to dominate pricing, and the exclusive access to global oil resources enjoyed by the major oil companies in the 1950s and 1960s has broken down.

Changing Scale Economies

Time and technology have changed the circumstances that led Frankel to conclude that vertical integration was optimal. In his time, the Seven Sisters controlled the volume of global oil production and price. Tanker fleets connected tidewater refineries to super giant oil fields in a smooth continuous operation. In that era, vertical integration was the key to profitability; centralization and scale economies were essential to match remote resources to dispersed markets.

Declining U.S. oil production and the rise of OPEC, however, had a major impact on industry structure. Declining U.S. production fragmented the industry. It meant smaller fields and higher costs, giving the advantage to independents. Rising oil dependence shifted deliveries to marine tankers. The industry discovered limits to scale economies, which ended many of the advantages of market concentration. Oil tankers no longer increased in size, stabilizing with the super tanker. Likewise, the sudden shift of property rights from the major oil companies to OPEC's national government resulted in an overnight involuntary dissolution of vertical integration. The next two decades saw a desperate, and largely fruitless, scramble by the major oil companies to return to bygone glories.

The High Cost of OPEC's Administered Prices

Chapter 4 described the birth and development of the oil commodity market and noted its success was based largely on an accident of timing. An industry dominated by vertical integration has little need to procure supplies on a daily basis or manage price risk. Luckily for NYMEX, the exchange launched oil trading just as the major companies lost control of OPEC oil. The shift in ownership, however, need not have meant the end of vertical integration. As previously noted, OPEC's national oil companies could have integrated downstream. Likewise, the majors could, and did, explore for and develop oil in other regions. It turned out, however, that the costs of both activities were high and in many cases much higher than expected.

With a few exceptions, OPEC's national oil companies lacked the expertise to integrate downstream. In general, the numbers of employees in refining and marketing are much larger than in the upstream sector of the oil business for the types of fields controlled by OPEC's members. Typically, refining is a low-margin business and profitability depends on having an efficient product distribution and marketing system. To maintain margins, companies establish brand names and compete through advertising and other public awareness programs. Management must speak the local language and understand local customs. All of these factors were an inhibition to OPEC's national oil companies. Since the upstream sector was by far the most profitable, it made sense for NOC's to simply market their output as a commodity and leave the more costly and complex task of refining and marketing to established companies.

The IOC's search for replacement crude oil had mixed results and a string of disappointments. The sudden loss of exploration rights within OPEC left them with a thin portfolio of exploitable projects. Their traditional focus was to search for and develop giant oil fields. These were the projects in which their engineering expertise and financial strength could be most effectively wielded and where smaller companies did not have adequate scale to compete. Consequently, it was difficult to shift investment focus away from the Middle East and its low-cost resources. Early on, the IOCs recognized the deepwater prospects in the Gulf of Mexico and offshore of Western Africa and Brazil, but the technology was just emerging. The Arctic in Alaska and Canada seemed to hold great promise, but as it turned out there were no major oil discoveries after the Prudhoe Bay and Kuparuk oil fields.

OPEC's tepid interest in refining and the IOC's sudden crude oil deficit left a large gap in the market. By 1978, the transition from company to government control over OPEC oil reserves was more or less complete. ARAMCO was still tacitly under the control of its original owners, but the Saudi government made important decisions on investment, production, and pricing. In spite of this, the IOCs continued to procure crude oil from their traditional sources, despite the fact that the terms of trade had changed dramatically against them.

The initial impact of the 1973 to 1974 oil shock was higher prices, rather than supply insecurity. The Iranian Revolution and accompanying supply dislocations five years later changed that. As the largest member of the Iranian Consortium, BP had been the most dependent on Iranian crude oil and suffered the largest shortfall. ARAMCO partners had the opposite problem; Saudi Arabia increased production but restricted resale. The kingdom issued the "Yamani edict," which prevented the ARAMCO partners from reselling Saudi oil at anything other than official prices. As consequence, it was to the advantage of ARAMCO companies to refine Saudi oil rather than resell it. They announced that they would not renew crude oil sales contracts, inducing a new round of anxiety and creating serious supply problems for many independent refiners.

As explained, the oil industry is one of continuous flow; thus, scheduling is crucially important. In the world oil market, this manifests itself as a steady stream of oil tankers going out and coming in. Tankers make a continuous loop from oil exporting to oil importing countries and transit time between the two can be as long as 60 days. Often the oil stored in transit exceeds the working inventories of oil ashore. The logistics of the flow is precise. Oil flowing from production wells goes to storage facilities at marine terminals, but the capacity is limited. If tankers do not show up as scheduled, the tanks overfill and production shuts down. Likewise, if too many tankers show up they could stand idle for days with the possibility of creating shortfalls at consuming nations' refineries. Oil tanker scheduling is not a precarious balance (unlike power grids) but it does require planning and organization. Thus, the operation is smoothest when there are stable relationships between buyers and sellers. In other words, longer-term contracts (or vertical integration) play an important role in managing oil's continuous flow from the wellhead to the gas tank.

The sudden collapse of Iranian oil production in 1978 created logistical confusion and, more importantly, refiners cut off from traditional supply sources panicked as they faced the prospect of empty holding tanks and half-filled stills. The consequences were frequent price spikes for the limited number of spot cargos. Much higher spot prices, in turn, tempted many OPEC members to abrogate term contracts in favor of spot sales which further destabilized the market as tankers rerouted and negotiations faltered. OPEC tried to discourage its members from selling in the spot market, but the incentives were powerful and all too often a traditional customer found that scheduled cargo pickups had been preempted by a fleet-footed broker.

Figure 7–1 is a sample of various crude oil purchase prices in 1979 and 1980. The scatter diagram relates crude oil by type or source (color and symbol differences) to f.o.b. price and date of purchase. During this time period, the Energy Regulatory Administration (ERA) of the Department of Energy collected price information of each cargo of imported oil in order to manage the price control program. Between January 1979 and the end of the program in January 1981 over 5,000 cargos of crude oil were imported. The U.S. imported oil from every major exporting country. The data submissions by importing companies did not make a distinction between spot and term prices, but that is generally evident by the prices paid.

The ERA-51 data reveal a number of interesting features of oil purchases during that time-period. First, there are a variety of crude oils in the chart, with significant differences in quality and location. The cheapest crude oils were heavy and sulfur laden. The priciest were light oils, close to markets. Secondly, oil purchases did not divide neatly between term and spot categories. There were many intermediate deals. For example, refiners often had to purchase several cargos at spot prices in order to obtain a term contract at official prices. Many of the cargos in the chart were for the resale of oil in transit by a third party, rather than at the oil-exporting country's terminal.

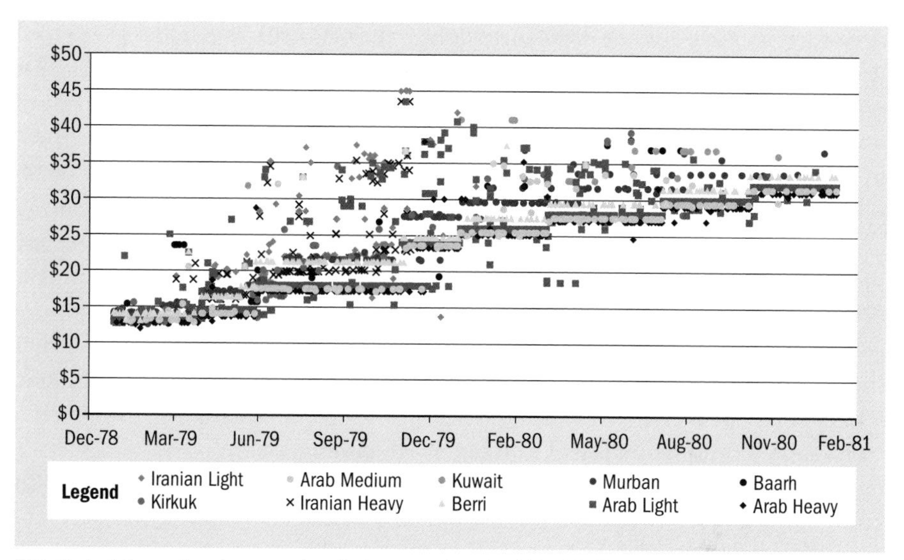

Fig. 7–1. Prices of various crude oils imported to the U.S. 1979–1980. *Source: ERA-51*

The vast bulk of crude oils imported sold at official prices. This is evident by the close-knit observations toward the bottom of the chart that form a band of prices. There were a few odd lots selling below official prices, but by-and-large, if prices deviated from official prices they were higher. What might be a surprise for some observers is that there was a vast range of prices at any one time. This generality holds for even for crude oils of like kind and quality. Spot prices reported by the trade press exaggerated the differential because they reported only the highest of the spot transactions.

The type of market that these data describe is not supposed to exist and certainly not for a period of over two years. Economists generally believe in the law of one price (LOP). The LOP asserts that given reliable information, low bargaining costs, and modest enforcement costs, arbitrage will cause price convergence. At the heart of this assertion is the proposition that buyers and sellers will act rationally in their own self-interest. It can be (and often is) argued that not all consumers act precisely to their own advantage. That is a hard argument to make about oil traders, unless they have had one-too-many at the petroleum club. Furthermore, these cargos were worth millions of dollars, so it is unlikely that irrationality was at the heart of the decision process. Instead, it is obvious that bargaining and enforcement costs were high and information sketchy—in short, high transaction costs. Anyone active in that market at that time will confirm this observation.

At this point, it is worth stepping back and once again reviewing the OPEC price methodology. The cartel was supposed to set the Saudi Light official price, and let Saudi production adjust to balance the market. That mechanism, however, depended on huge volume adjustments, given the seasonal and emotional components of the oil market. If demand was too high and stretching capacity, or if it was too low and endangering the cash flow of the kingdom, the price mechanism would slip into disarray. As it turned out Saudi production capacity was too small in 1979 to 1980 when the market spiked and too large from 1985 to 1986 when it collapsed. During these periods, the trading and logistical costs of procuring crude oil were extremely high. The high cost of procuring oil might have pushed the industry back to vertical integration were it not for the decades of resentment built up among OPEC members. Instead, a far more efficient system of futures and commodity trading emerged. The revised system achieved logistical flexibility through market flexibility, specifically price flexibility. It is an old law of economic theory that sellers can fix supply, or price, but not both. That is why the OPEC cartel shifted from fixing prices to fixing production quota.

It is important to note that trading costs, as a key element of transaction costs, have changed over time as energy markets have matured. Figure 7-2 illustrates the trend of trading costs, measured as the difference between buy and sell prices, for the purchase and sale of three key energy commodities. This calculation effectively measures of liquidity—the lower trading costs, the more liquid the market. As noted, Arabian Light crude oil was for many years the marker price set by the OPEC cartel and is the largest volume crude oil stream in the world. Brent crude oil is the marker for the ICE futures market in London. Henry Hub is the most important natural gas trading point in North America and is the marker for natural gas futures trading in New York.

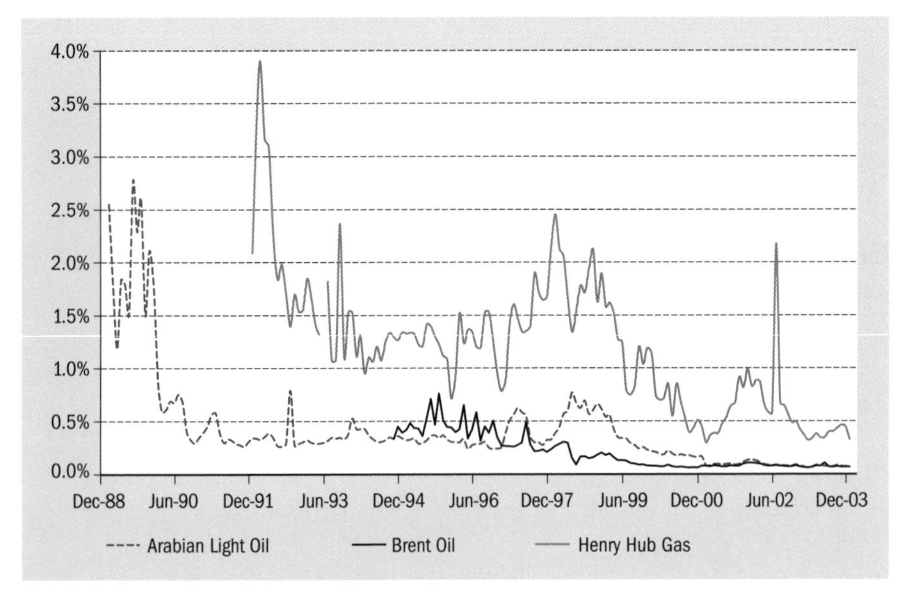

Fig. 7-2. Change in trading costs. *Source: Reuters*

The difference in bid and ask prices has dropped for all these commodities, as their markets matured. Figure 7–2 illustrates a steady drop for Arabian Light crude oil, except in 1998 during the Asian financial crisis. Since the late 1980s, crude oil's trading costs by this measure have been exceptionally low. In contrast, natural gas trading costs are higher. This is because gas has a fixed infrastructure delivery system, which regionalizes prices. The California energy crisis of 2001 was a prime exception to the downward price in trading costs in natural gas.

Traders use futures markets mainly to manage price risk; market efficiency is the consequence not the purpose. Surprisingly, there was an extraordinary side effect to the market transformation; a shift of focus no one expected. Once an efficient trading system was in place, companies in the industry could focus on what they do best; they could specialize. The shift greatly reduced the incentive to integrate vertically, undermining the historic structure of major oil companies.

The Elements of Specialization

Broadly, the petroleum industry divides into an "upstream" and "downstream" sector. Primary upstream activities, as the name implies, are exploration and production (E&P). Because explorers often find oil and natural gas together, the industry includes both in the upstream sector. The primary activities for the downstream sector are refining, distribution, and marketing. A variety of transportation options link upstream and downstream activities. Oil ships in pipelines, marine tankers, rail cars, or tank trucks. Natural gas moves by pipeline from the field to a distribution system. For intercontinental commerce, natural gas producers liquefy the gas and ship it by marine tankers as liquefied natural gas (LNG). Each of the primary activities involve a whole set of specialized support services and activities. For example, producers normally contract for drilling rigs, rather than keep own their own. Most refiners have cogeneration plants that produce electricity for the refinery's use and resale, and firms that specialize in the electricity industry often contract to build and operate these plants.

The fundamental questions for the study of the oil industry's structure are how are all the activities linked together and why do the majority of companies take on a particular form? At one extreme is a vertically integrated arrangement in which a company finds, produces, transports, refines, and markets oil as one continuous operation. At the other end of the spectrum are companies that operate only in one sector—oil producers, transport companies, refineries, and marketers. Put in the jargon of the deregulation movement: is it a bundled or unbundled industry? An unbundled industry allows companies to specialize; to focus on what they do best. On the other hand, vertical integration allows systemized planning and reduced risk. Critics of the industry have complained that vertical integration leads to market concentration and less-than-competitive pricing. However, specialization may leave the industry without long-term planning and investment, driving up price volatility. Oil and gas production from the Arctic, deepwater, and other frontier resources requires a long planning horizon. So far, such development has been the province of major integrated oil companies who have the experience and patience to proceed despite volatile prices.

The organization and management of the upstream and downstream sectors are very different even though continuous flow knits them together. The skills required to successfully prospect for oil are largely unrelated to the skills required to run a modern refinery and market its output. The synergy of these activities is more subtle. Crude oil has very little direct value; it must be refined into petroleum products to meet industry-wide specifications for use in cars, furnaces, jet planes, and so on. Producing crude oil is just as capital intensive and costly as owning and operating a refinery. If the oil producer does not have a market, or the refinery a source of supply, their investments may stand idle with mounting losses.

The petroleum industry encompasses a trade-off between planned and coordinated flow from the field to the consumer under the management of one entity, as compared to the benefits of focused specialization in one aspect of the production and consumption cycle. The impact of the trade-off between specialization and coordination has waxed and waned over the industry's history and depends critically on the efficiency of oil markets and the ease with which private companies have access to the most prolific resources.

Figure 7–3 presents a flow diagram of the industry, including some key aspects of its activity. Activities associated with the upstream segment are in white. Transportation and storage efforts are shaded in medium gray, and the downstream sector, refining and marketing is shaded in dark gray. Along with the major sectors of activity, figure 7–3 includes examples of many specializations that support the oil industry. Key components of E&P include leasing, exploration, field design, infrastructure support, and development. If the field is offshore it will also include platform construction and marine services support. For refining, the key activities are crude oil acquisition and trading, risk management, ensuring product specifications, and safety training. Marketing involves creating a brand name, locating distribution outlets, managing independent station operators, managing credit and contract issues, and so on. Transportation and storage supports each of these sectors and safety is essential for all segments of the industry.

Upstream

E&P is the sector of greatest interest and has usually been the most profitable. However, a major consequence of maturing oil fields is the increase in development complexity. Oilmen of the early twentieth century needed only two basic resources, the chutzpa to obtain mineral rights and a drilling rig. In contrast, major field development today entails a wide variety of support activities. The

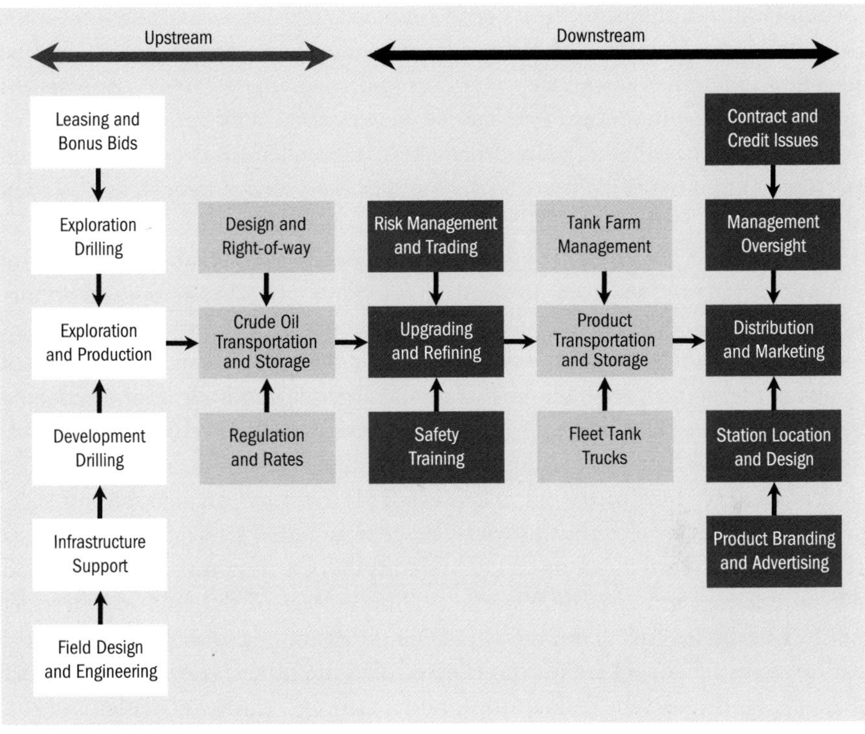

Fig. 7–3. Petroleum industry activities

Petroleum News Directory now lists 98 categories of support; everything from underwater welding to environmental engineering. Recently, Alaska's production manager for ExxonMobil, commented that more than "150 people from more than 30 companies in Alaska are working to progress drilling and development activities for the Point Thomson field." (*Petroleum News* 2008) (The Point Thomson field is east of Prudhoe Bay on the North Slope; it is primarily a gas field, but also contains significant volume of natural gas liquids. Although discovered in 1977, Exxon postponed Point Thomson's field development, awaiting the construction of a natural gas pipeline.)

Exploration geologists have surveyed most of the globe for its oil and gas potential. Indeed, a good deal of present exploration begins with geological maps and analyses of the early twentieth century. Generally, geologists know the location of potential oil and gas structures; uncertainty surrounds the technology, economics, and politics of development. For most companies, it all boils down to a simple issue, is there a profit opportunity? This question has three components: Is it possible to obtain exploration and development rights? Is a reliable technology available to produce this type of resource? Is the deposit likely to be sufficiently prolific to cover the capital and operating costs necessary to produce oil or gas?

Before any exploration can begin, companies must consider taxes, royalties, bonus bid payments, participation requirements, environmental offsets, and all the other external factors that affect cost and financing. Many companies do such analysis in-house, but a variety of contractors offer consulting services and reports that compare relative fiscal terms. Smaller E&P companies often specialize in one region where they have expert knowledge in both the geology and contracting terms.

Most onshore oil wells in North America are on private land, with mineral rights leased by one or more production companies. Royalty terms and conditions vary depending primarily on knowledge about the resource before drilling commences. Typically, royalty rates are around 15%, but they can be higher. For example, it was well know that the Wilmington oil field in California extended offshore with the City of Long Beach holding the mineral rights. Long Beach held an auction for the field development. The wining consortium of Texaco, Humble, Mobil, and Shell (THUMS) bid a royalty rate of 99%, less production costs. In this instance, the companies all had refineries close at hand and their motive was to secure a feedstock that minimized transport and handling expenses.

Oil companies have explored almost all the private land in North America that has potential for oil and gas production. On the other hand, a large number of public lands, notably federal offshore acreage and tracks in Alaska and the Rocky Mountain states have yet to be leased or drilled. In the U.S., the Mineral Management Service (MMS) of the Department of Interior is responsible for oil and gas leasing. Prior to an auction, the MMS does a preliminary assessment of the expected reserves on each track, so that bids must meet a minimum standard. The U.S. also uses a bonus bid system in which companies bid up-front cash payments to obtain a lease. Once production begins, there is also a royalty obligation. Leasing terms usually include minimum performance standards. For example, if the winner does not drill within a set timeframe, the lease expires.

Outside North America, Australia, and parts of Europe, most mineral rights reside with the national government. This is a simple, but often overlooked constraint, especially since most future supplies will come from these sources. Historically, oil companies negotiated directly with the host country's government, striking a deal on drilling rights and royalty payments. The foreign oil company did all the work—exploration, field development, and marketing—retaining control and equity rights to the oil. Over time, however, countries with significant oil export potential developed their own expertise in the form of a Petroleum Ministry and/or a National Oil Company. As expertise in the host countries grew, and the OPEC revolution unfolded, the host countries took on more and more industry tasks. Today, oil-exporting countries use two main types of arrangements that allow foreign companies to develop their resources: Straightforward service contracts for specific expertise to assist in developing or maintaining field production and/or production-sharing agreements (PSAs) where the foreign oil company will carry on the primary tasks but share production with a local private or government-owned oil company.

Downstream

In the 1950s and 1960s, crude oil supply was abundant and the primary task of crude oil traders was to keep refineries full. The periods of crude oil scarcity and far greater price volatility since then have shifted the focus of traders and refinery managers. Successful refineries are substantially more complex than in earlier decades; they run lower quality crude oils and produce far less heavy fuel oil. There is little or no waste; surplus gases are used to produce electricity and residual oil is converted by coking or cracking to light petroleum products. In order to optimize refinery runs, traders have to adjust the feedstock input depending on the prices of various crude oils and the value of products they produce.

Typically, refiners have a number of crude oil sources, which constitute a "base load." Specialized refiners typically acquire crude oil using long-term contracts. Integrated companies depend on crude oil from their own proprietary sources, although they may enter into exchanges to save transportation or other logistical costs. For example, before BP acquired refining and distribution systems on the U.S. West Coast, it had a substantial surplus of North Slope crude oil. It sold much of the oil outright, but it also exchanged Alaska oil delivered in the U.S. for foreign crude oils delivered to its refineries around the globe.

Refineries typically top off their base load supplies with varying amounts of heavy or light crude oils depending on the season and relative prices. Traders often complain that they spend 90% of their time managing 10% of the supply. However, refinery margins are seldom robust and the final portion of the feedstock can make or break a refinery's operation. Linear programming models guide refineries crude oil acquisition strategies. Given product prices, the models determine which crude oil or combination of crude oils maximizes the refinery margin. Although it is possible to generalize about the components of a sophisticated refinery, each has a unique design. Consequently, the models that seek to minimize cost and maximize revenue are specific to a refinery.

Safety is a major concern everywhere in the oil industry, but particularly in refining. Obviously, the combination of heat and highly flammable material is dangerous anywhere, but the proximity of refineries to population centers compounds the risk. It should not be surprising that refiners spend a good deal of staffing time and training managing a refinery's safety.

Unlike the E&P sector, refiners integrate many of their activities within a firm. The exception, of course, is when refineries are being constructed or modified. In those instances engineering and construction firms are used to design and build the facilities. This is akin to the exploration and field development in the E&P sector. Once a field is developed, operating the facilities requires consistent management over time rather than the coordinator of a wide variety of specialized skills. Aged fields, however, require the application of specialized skills. Refineries require maintenance, but do not require upgrading to retain capacity. In any case, activities that are highly specialized and indirectly related to refining are contracted to specialized firms. Power generation, marketing, linear programming, and so on are the sorts of activities likely to be external to the refinery company.

In contrast to E&P and refining, petroleum marketing is technically simple, but contractually difficult. Retailing is the interface between brand name oil companies and consumers. Consequently, it is the most contentious and the focus of much public attention. Suffice to say that product marketing is a complex combination of integrated operations and contractual relationships. Although the balance of the two approaches has varied over time, there has always been company outlets intermixed with independent marketers (which the industry often refers to as *jobbers*). The balance between the two is dependent on the types of joint product offered by retail outlets, the opportunities afforded by superior locations, and the complications of managing remote operations.

The Modern Theory of Transaction Costs and the Firm

The theory of the firm

Nobel laureate Ronald Coase (1937) asked a question central to the study of economic institutions: "If markets are the most efficient means to allocate resources, why do firms exist?" Coase answered the question by explaining that if the transactions cost of using a market was higher than the cost of an internal resource allocation, firms would expand. Any study of the oil industry must address this question, because the energy sector is where many of the world's largest corporations operate. The elephantine size of energy companies is not new; it has been a characteristic of the industry since its early foundation. Only two firms have appeared systematically in *Fortune's* list of the ten largest firms: Standard Oil of New Jersey (once the Standard Oil Trust and now ExxonMobil) and General Electric. John D. Rockefeller founded the Standard oil companies and many insiders believe his personality mirrors current operations and decisions.

The problem of asset specificity

Since Coase's (1937) famous article on the theory of the firm, "transaction costs," as information, bargaining, measurement, and enforcement costs—are the costs of using the marketplace. The basic notion is that when these costs are high, firms internalize resource decisions and avoid trading. One branch of transaction cost economics, identified with Oliver Williamson (1985) and Klein,

Crawford, and Alchian (1978), has analyzed the impact of these costs, particularly bargaining and enforcement, on a company's organizational structure and contractual arrangements. Bargaining and enforcement costs typically rise, the more specific or special the traded commodities or assets.

Some assets are flexible and some are not. Farmers use tractors for a variety of tasks including ditch digging by adding a backhoe. On the other hand, contractors use specialized trench diggers to dig ditches more quickly and efficiently. Such a tool, however, is useless for plowing a field or loading hay. A good share of the capital equipment used by the oil industry is customized design and permanently installed at specific oil fields, transport corridors, or refineries. In short, the investment has little or no alternative uses.

Asset specificity creates asymmetrical problems for markets. Before design and installation, there can be a great deal of competition. That is, a large number of qualified companies could be capable of constructing and operating the facility. Likewise, such companies could bid on a variety of construction projects around the globe. Once built, however, competition narrows and the relationship may become one-on-one. For example, consider a pipeline from an oil field to an ocean terminal. Before construction, any number of qualified companies could build the pipeline. Once built, however, the pipeline owner has monopoly power over the oil producer—shipping costs using trucks or rail could be much higher. The analogy does not end there, however, because if there is only one oil field, the producer has a monopoly too. The pipeline needs to ship the oil if it is going to have any revenue. The problem can be neatly summarized as follows. Ex-ante construction there is competition, ex-post there is bilateral monopoly.

Ex-post bilateral monopoly, monopoly, or monopsony opens the door to opportunistic behavior. Once a project is finished, one or both parties to the deal may see an opportunity to extract "rents" from the other by threatening to curtail or shut the operation down. Professor Klein refers to this somewhat tongue-in-cheek by a less-technical description: "the holdup."

Economic theory identifies the essence of the difficulty, but, as usual, there are real world complications. Figure 7–4 is a matrix that describes possible outcomes following construction of specialized assets.

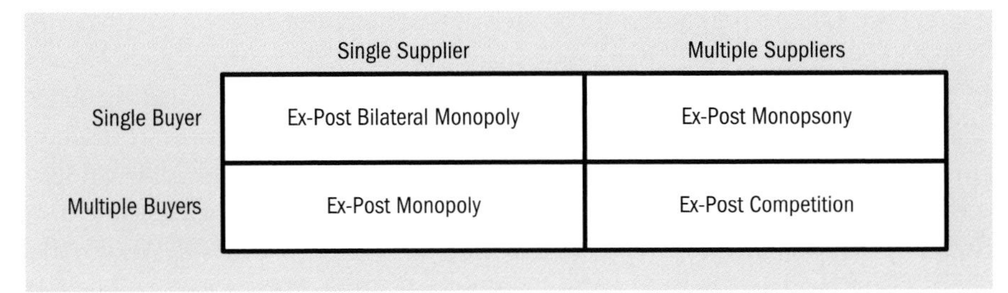

	Single Supplier	Multiple Suppliers
Single Buyer	Ex-Post Bilateral Monopoly	Ex-Post Monopsony
Multiple Buyers	Ex-Post Monopoly	Ex-Post Competition

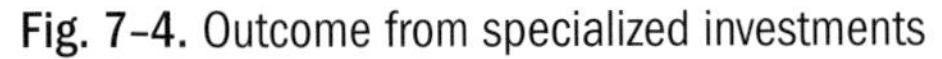

Fig. 7–4. Outcome from specialized investments

Examples of industry structure

The top left cell of figure 7–4 describes the outcome most often envisioned by economic theorists, a single seller and single buyer. Faced with ex-post bilateral monopoly, what do companies do? There are two fundamental choices. The buyer could enter into an enforceable long-term contract with the seller or acquire the asset and operate it as a vertically integrated enterprise. The choice will depend greatly on the company's confidence in the rule of law, i.e., the enforceability of the contract, the ex-ante bargaining power of the parties, and the optimal degree of specialization.

Historically, the international oil industry had an incentive to integrate vertically, because at the time the industry-developed companies could not count on contract enforcement across multiple countries. To be secure, the companies had to handle the oil within the same organization through the entire value chain. There was also an important side benefit reinforcing the industry structure, once it was in place. Because OPEC members had no marketing channel or alternative outlet for their oil, it was suicidal to expropriate foreign producers' assets. Likewise, marketing operations and refinery investments were secure because only the major oil companies had access to cheap foreign crude oil. The panic over supplies that followed the Arab oil embargo of 1973 shifted the perception. By the end of the next decade, the oil companies' equity interests in OPEC resources had effectively vanished.

The cost of negotiating crude oil purchase contracts also had an impact on the North American oil industry, and encouraged vertical integration. As explained earlier, the diversity of crude oil means that for optimal results refineries are designed to process specific crude-oil streams; substituting other types of crude oils results in lower volume runs or excessive production of low-valued products. An example will help clarify this point. Originally, the Getty refinery in the Bay Area of California connected to Getty's Kern River oil field by pipeline. This crude oil is heavy, but low in sulfur. Getty sold the refinery, but not the oil field to Tosco. Instead, it entered into a long-term sales contract with Tosco. Later, Texaco acquired Getty and the purchase included the Kern River field. Other heavy crude oils with high sulfur were available to Tosco, but could not be as profitably refined. Refurbishing the facility to be more flexible would have added to capital cost. The lack of flexibility put Tosco at a bargaining disadvantage in the 1980s when renegotiating its purchase contract with Texaco.

Vertical integration may be self-reinforcing, in that it increases costs for those that do not integrate. The more integration, the lower the number of trading opportunities in the market, and the greater the difficulty of optimizing refinery runs. This arises in the first place because there are simply fewer crude oil streams available. The specific asset problem compounds the problem; with less active trade, the instances of opportunistic behavior are likely to increase. Once in place, it takes a major shock, such as the OPEC revolution, to dislodge vertical integration.

Pricing is not the only issue regarding specific assets, access and control can be just as important. Oil and gas producers often combine to sponsor a pipeline. Competitors in one sphere cooperate in another. Ownership of the Trans-Alaska Pipeline System (TAPS,) for example, is more-or-less, proportional to the producers' original interests in the Prudhoe Bay oil field. The Sudanese oil company, CNPC, and other equity participants in Sudan's oil fields sponsored and built the notorious Sudan oil pipeline that connects fields in the far south of the country to Port Said in the North. In both of these examples, a very costly transportation system was necessary to produce and move the oil. In both cases, the oil producers found it in their interests to jointly own associated pipelines, rather than contract to a separate party. Joint projects resolve the access issue for the sponsors but not for third parties that might make new discoveries. Without regulation, control of key transportation systems can constrain competition. On the other hand, regulations themselves can inhibit pipeline projects, because they reduce the owners' control and constrain revenue.

In contrast to the oil market, most cross-border trade in natural gas has developed using long-term contracts, rather than integrated ownership. The reasons include the improved reliability of contract enforcement, the economies of scale of natural gas distribution systems, distinct patterns of specialization and an objective basis for pricing.

Other than a few isolated cases, cross-border trade in natural gas did not develop until after World War II, when the environment for reliable trade had vastly improved. The Bretton Woods agreement, the General Agreement on Trade and Tariffs (GATT,) the creation of the European Common Market, and extensive bilateral treaties created a system of contract enforceability that did not exist in the earlier era when the oil industry developed. Without these institutions, the risk of building inflexible delivery systems would have been too great.

The key to the natural gas industry is its highly specialized pipeline distribution network. It is common to view gas and electric distribution systems as natural monopolies, with obvious consequences for pricing and behavior. In most jurisdictions around the world, gas systems are either publicly owned or regulated. If a private company owns and manages the distribution system, exclusive franchises are normally only granted on the condition that there will be regulatory oversight. Historically most urban gas distribution systems began with gas manufactured from coal (referred to as town gas). In North America, the industry linked town gas systems to domestic natural gas fields beginning in the 1930s when new technology allowed the construction of high-pressure gas transmission pipelines. In contrast, Japan and Europe depended on town gas until 1960s. It was difficult for foreign natural gas to displace coal gas. Domestic gas distribution companies were obviously unwilling to abandon their franchise. In most cases, natural gas supplies were too distant and therefore too costly to compete, particularly without access to the distribution system.

The 1959 discovery of the Groningen gas field in the Netherlands was a key event, because it identified the potential for North Sea oil and gas discoveries that were to follow and because it set the contracting pattern for cross-border gas sales. The field is the largest in Western Europe and 1959, the Dutch calculated that its reserves were greater than domestic consumption could absorb. (At the time, there was an expectation that nuclear power would be cheap enough to displace residential and commercial gas uses and render the gas reserves worthless.) Consequently, the Dutch chose an aggressive marketing strategy to neighboring countries.

Unlike oil, natural gas development was constrained by the high cost of ocean tanker shipping. The technology to liquefy natural gas had been available since the nineteenth century, but not proven commercial until the 1940s. The first shipments of LNG was made in a converted Liberty ship from Louisiana to the UK in 1959 and the first continuing arrangements were made between Algeria, the UK and France in the 1960s.

The discovery of natural gas in Alaska's Cook Inlet opened up the possibility of U.S. gas exports to Japan as LNG. Gas exports were acceptable to the U.S. because Cook Inlet gas was far in excess of local Alaska demand and the U.S. was already importing surplus gas from Canada. In 1967, the Federal Power Commission (predecessor to FERC) granted Phillips 66 and Marathon an export license for Cook Inlet liquefied natural gas and exports began in 1969. The producers built the liquefaction facility and the buyers, Tokyo Gas and Tokyo Electric Power Company, took responsibility for transport and re-gasification.

Contracts for cross-border gas sales were of sufficient duration to recover capital on the extensive facilities. The parties involved were creditworthy companies, with well-established records of accomplishment. For example, an early contract for Cook Inlet LNG was between Tokyo Gas as the buyer and Phillips as the producer. National governments backed the arrangements, adding to security. These factors combined to ensure enforcement of contract terms. Natural gas also had an advantage over oil for long-term contract pricing, in that the parties could construct an objective index of competing fuel costs that would be fair to all. The theory being, that a rise in the price of heating oil would increase the competitiveness of gas and vice versa.

The structure set by the first contractual arrangements for cross border pipelines in Europe and liquefied natural gas projects in Asia set the pattern for the industry. As gas consumption increased, Japan's utilities contracted for LNG imports across the region from Brunei, Indonesia, Malaysia, Qatar, Oman, UAE, and Australia. Companies like Royal Dutch Shell and Chevron produce and liquefy the gas paying royalties, taxes, and so on to the host governments. Likewise, the region's strong economies, South Korea, Taiwan, India, and China, have built gas distribution systems and developed gasification facilities. To make the industry work, it took the expertise and technology of producing companies and the creditworthiness of large gas distribution companies, linked by enforceable long-term contracts.

In Europe, countries with natural gas resources available for export—the Netherlands, Norway, and Russia in the North and African countries Algeria and Libya in the South—connected to the consuming nations in central Europe by a series of high-volume gas transmission pipelines. Often these pipelines pass through one or more transit corridors in other countries. The Maghreb-Europe gas pipeline (MEG) is representative of the European commercial structure. Completed in 1996, the pipeline links the Hassi R'mel gas field in Algeria with the Spanish and Portuguese natural gas grids. Sonatrach, the Algerian national oil company, owns and produces the gas and constructed and owns the Algerian portion of the pipeline. The pipeline passes through Morocco, with that segment owned by the state government and operated by a joint venture of Spanish, Portuguese, and Moroccan companies. Enagas (the Spanish gas distribution company) and Morocco own the undersea portion. The balance of interests in the MEG pipeline is not accidental. The design minimizes the risk of opportunistic behavior, given its specific purpose and use.

In the natural gas market, pipeline ownership and control is only part of the complex economic relationship. Europe's natural gas remains sourced from just a few countries where national policies consolidate resource control into state-owned or controlled monopoly suppliers. This combined with the risks associated with the development of high-volume long distances transmission pipelines has necessitated long-term contracts, but, in turn, limits competition, and yields a clumsy system of pricing. There is, however, an exception just across the English Channel.

The present UK natural gas market is similar in structure to the North American gas market and is an excellent example of ex-post competition, as identified in the lower-right-hand box of the matrix shown earlier in figure 7–4. The grid is privately owned, but regulated, allowing third-party access. There are independent North Sea gas producers and competitive marketing to retail customers. There is also an active natural gas futures market operated by ICE in London. Gas prices are marked against the national balancing point (NBP) of the grid. The question naturally arises as to why the UK gas market structure and pricing differs so greatly from continental Europe. The reasons are primarily due to Margret Thatcher's effort to privatize much of the UK's nationalized industry, and the location and type of the resource.

Like the U.S., the UK has a variety of independent private gas producers feeding gas into the nationwide grid. The gas flows from multiple fields in the North Sea. Some of the gas is actually from Norway but landed in the UK because of the depth of the Norwegian Trench. Viewing the system as a whole, there are few bottlenecks or other barriers that would balkanize the market and, compared to Europe, transmission pipeline distances are short. This arrangement opens the door to supplier competition. Likewise, UK shifts in regulatory policy allowed open competition in the retail market with consumers free to choose their supplier.

The bottom left cell of figure 7–4 describes a situation quite familiar to oil and gas producers. It is frequently the case that a single pipeline links multiple fields to the market. Some of the linked fields may have potential for increased production. As explained, oil and gas pipelines have strong economies of scale, which leads to a natural monopoly. Quite obviously, most are specific assets too, tied to one location with limited alternative value.[2] If there is no regulatory oversight and the pipeline owner is also a producer, rivals may have difficulty obtaining pipeline access. Third-party access is often a bigger issue than shipping rates to competing producers. It is also a concern to regulators and policy makers, because shipping discrimination can have a detrimental impact on field development in that it inhibits further exploration or blocks low-cost fields from the market.

Controversy over oil pipeline access was a contentious issue in the U.S. for decades. Once again, the issue revolves around Standard Oil, which at the end of the nineteenth century controlled approximately 80% of the country's refineries and pipelines. That changed in 1906 with the passage of the Hepburn Act that defined oil pipelines as "common carriers" under the Interstate Commerce Act of 1887. The Interstate Commerce Commission regulated rates and access until 1977 when it was replaced by FERC. Without getting into the detail, the essence of U.S. regulation in the early years was to assure equal access for shippers at cost-based rates. Companies wishing to ship oil would nominate their expected volumes and if the total exceeded capacity, the pipeline company would cut back each shipper by an equal percentage. In recent decades, this has shifted to a "light-handed" regulation, where FERC allows market-based rates, as long as the pipeline owner can demonstrate the absence of market power. The shift in regulatory philosophy reflects the decline in U.S. oil production, which has resulted in substantial unused capacity.

Although many view a pipeline as a standalone monopoly, it is also a component of joint production, i.e., part of field development or a refinery operation. For example, assume an oil field producer sponsors a pipeline designed to match the field's output. In some cases, U.S. regulations will allow a portion of a competitor's nearby production on the pipeline. This can back down oil production from the sponsor's field and the cost in lost production revenue can be far greater than that recovered from pipeline transport fees. (Recall that both field and pipeline costs are almost entirely fixed and do not drop with lower volume.) Likewise, a refiner may not be able to process at full capacity, because regulations require sharing a pipeline with a competitor. These simple examples, demonstrate that the difficulties of managing and regulating specific assets when they are unbundled are far greater than the ex-post bilateral monopoly case most commonly analyzed. It is often difficult for regulators to identify anti-competitive behavior, let alone make an efficient correction.

Specific assets can create an ex-post monopsony or its appearance, as noted in the upper-right-hand portion of the matrix in figure 7–4. A single refiner-pipeline owner connecting to a number of privately owned oil fields is a good example. In that case, the refiner could force crude oil prices down. This is, of course, the historical accusation made by independent producers against Standard Oil.

There are other interesting examples of monopsony behavior. Jonathan Stern (1998) refers to the large gas distribution companies in Europe as "gate-keepers." Cabot's LNG terminal in Boston has had virtually exclusive access to the New England gas market. Gas from the facility competes with pipeline gas flowing from the South. Cabot exercises market power, however, with regard to other LNG companies that might wish to market directly to New England, rather than use terminals in the Gulf Coast and a long pipeline haul north.

Asset Specificity and the OPEC Revolution

Most of the specific asset examples just described relate to natural gas, rather than oil. This would not have been the case in the early twentieth Century, when Standard Oil's control of rail and pipeline transport supported its market power. Since that time, the world of oil has changed in ways that modify the analytical conclusions made decades ago. A number of factors have offset the industry's scale economics and asset immobility. Population centers in developed countries have shifted, with coastal areas enjoying the greatest growth. Most refineries are at or near tidewater. A good share of oil moves by marine tanker and barge, which does not exhibit the same economies of scale as pipelines and provides significantly greater flexibility. As noted, U.S. oil production has declined, resulting in surplus capacity of many pipelines. In any case, FERC regulates interstate crude oil and product pipelines as common carriers, which provide open access to would-be shippers. The problem of specific assets is much more obvious in gas and electric markets than in the oil market. In short, OPEC picked a good time to revolt and following that revolt, the oil industry has shifted away from vertical integration as its players have become more specialized. This might not have been possible earlier in the industry's history.

Nonetheless, some aspects of the specific asset problem remain in the oil market. The fourth, bottom-right, cell of figure 7–4 identifies circumstance where after specialized investments have been made there is competition—a sufficient number of buyers and sellers to stifle opportunistic behavior. Gasoline service stations in remote locations or in exclusive neighborhoods may on occasion take advantage of their single seller status to charge excessive prices, but their market power is limited. Even with ex-post competition, there is an uneasy interdependency of buyers and sellers. When both buyers and sellers must make fixed cost investments in specialized equipment to produce or use a "component" or standardized product, demand and supply schedules are highly inelastic. If the component product sells on a spot basis, it can lead to extreme price volatility (Van Vactor 2004, p. 30). In other words, the choices consumers have made to invest in energy-using equipment that require a specific type of fuel or energy source leaves them vulnerable to disruptions and considerable price uncertainty. Likewise, the investment producers have made leaves them vulnerable to periods of market stagnation and prices that result in little or no return to capital. This vulnerability often expresses itself in the quest for "energy security."

The Role of Government

Most of the important features of the oil industry's structure have been set by a natural evolution of private-based institutions seeking to maximize profits; the traditional view of how and why firms are organized. Abundant supplies of crude oil in regions with low population density and easy access to tidewater allowed the IOCs to integrate from retail marketing and refining to crude oil production and transportation. In so doing, they constructed a global system of supply and distribution, which they then dominated for decades. Likewise, when OPEC broke the chain, commodity exchanges developed to take over the task of determining crude oil prices and managing risk; tasks essential for the efficient allocation of resources. As long as prices were low and stable, government's role in many countries was simply to get out of the way.

Outside of the English-speaking group of nations, however, government intervention in the oil industry has been deeper and longer-lived. A perception that the oil industry is a tool of Anglo Saxon countries to perpetuate commercial advantages and by traditional concerns about energy security drives much of the intervention. In the last decade, many governments have sought to own or control the oil companies that manage domestic resources and provide their hosts with vital energy supplies. High oil prices and the financial crisis accelerated this trend. The consequence, as described in chapter 8, is a complex industrial structure, with a mix of national oil companies, government ministries, corporations, and companies combining public and private ownership.

So, to return to this chapter's theme: does form follow function? The answer turns out to be quite complex. The oil industry's structure is due in part to the physical qualities of the commodity itself, in part to actions of governments around the globe, and in part to the accidents of history.

Notes

1. 1986 low world oil prices and a surplus of crude oil from Alaska drove California heavy oil prices down to a few dollars per barrel. The marginal cost of extracting heavy oils with steam is costly and as a consequence a large number of wells in the San Joaquin Valley were shut down. This proved, however, to be a costly decision. Without continuing production the reservoir cooled and in some cases production could not be revived after oil prices recovered.
2. Pipeline right-of-ways have been used for telecommunications. Pipelines are sometimes converted to alternative uses. They may be used to carry CO_2 or are converted from oil to natural gas.

ALTERNATIVES TO OIL

Competing Futures

Imagine a world in the future when oil is scarce and as precious as gold or silver. Peak oil advocates see that world emerging in just a few years. Others believe it to be decades (or centuries) away. However long the present oil age lasts, there is near consensus that it will eventually end, and replacements will emerge. Just as coal replaced wood, and kerosene replaced whale oil, some alternative or alternatives will replace the vital fuels used today.

It is not a new idea to think about the coming energy transition. Indeed, advocates of Eisenhower's "Atoms for Peace" initiative had a master plan in mind. The idea was to transition from fission to fusion nuclear power; then, massive fusion plants would generate electricity and, using electrolysis, split water into hydrogen and oxygen. Hydrogen would become the primary fuel, piped to homes and industry for heat and mechanical motion and used in fuel cells for transportation. The result would be abundant energy supplies; hydrogen and oxygen would combine, exhausting only water. Undoubtedly, cynics would brood about the "mist crisis" clogging highways and urban centers. Nonetheless, the early vision was an endless cycle of clean affordable energy and, by implication, no limits to industrial and economic growth.

In contrast, the most strident advocates of peak oil envision an outright collapse of industrial society. Other, less rabid, futurists see a return to a nineteenth-century society, with tapped-out suburbanites fleeing homes too large to heat and cars too costly to run. The migration of the twentieth century will reverse, tenement housing in cities will swell, and farmhands will return to the soil with horse-drawn plows, casting mechanization aside. Trolleys and trains will replace private vehicles, sweaters will replace central heating, fans will become popular as air conditioners rust in the junkyard, and cooking by microwave will become a true art form.

There are other modern, and far more optimistic, visions of a world without oil. Imagine tomorrow's suburbanite commuting to work in a lightweight electric car. The car itself could be fitted with pedals to supplement battery charging and save the commuter time by making the daily gym visit unnecessary. Solar cells, wind generators, and other forms of renewable energy could charge the car's battery when at rest. Advanced insulation and solar architectural design would shield even the largest homes from the elements. The heavy footprint left by hydrocarbon use in the twentieth century would slowly disappear and humankind would live in harmony with the natural world.

These future visions are just that: imaginary visions of a future world often biased by the preset notions of the visionary. The path that phases out today's conventional fuels and the alternatives that will replace them is largely unknown. As Harold McMillan remarked when asked what governed his decisions as Prime Minister: "Events, dear boy, events." Events, and new technologies, are by their nature unpredictable. Although the distant future world—the world without oil—remains a blank page, facets of the forthcoming energy transition are emerging. Investments made today will shape the energy system of tomorrow. Nonetheless, energy transition constitutes a classic chicken-and-egg problem. Will supply follow demand for new types of energy products or will demand follow the availability of supply?

The Path to an Alternative

Lessons from the past suggest two quite different models for energy transition. First, was the switch from whale oil to kerosene for lamp lighting. Second was the switch from coal to petroleum as the dominant fuel for transportation and other non-stationary uses. The shift from whale oil to kerosene was comparably straightforward. Distribution changed little and existing lamps could burn either of the oils. Whale oil production increased substantially through the 1840s, peaking at around 15 million gallons per annum at the end of the decade. By 1860, however, production was down to around 10 million gallons and, by 1870, under 5 million. The drop in production was primarily due to resource depletion; whalers had spread out all over the globe, all the way to the Arctic and the Antarctic. As the intensity of the hunt increased, the number of whales declined, and the price of lamp oil went up, encouraging even more costly expeditions. Interestingly, the industry was largely located in the U.S.; at the industry's peak in 1840, over half of the world's 700 whaling ships were home ported in New England. There were nuances to the market; oil from sperm whales was particularly valued because it had a minimum odor when burned. Turning to smelly petroleum was a sacrifice for some.

Whale oil production had already begun to decline in 1858 when Colonel Drake made his famous discovery of crude oil in Titusville, Pennsylvania. The significance of that discovery was not just the drilling innovations that allowed the tapping of underground crude oil reservoirs. Rather, by 1858, it was clear that distilled crude oil's principal product, kerosene, could compete with whale oil in the illumination market. In other words, crude oil suddenly had real commercial value, created by the collaborative efforts of an entrepreneur, George Bissell, and a chemist, Professor Benjamin Silliman (Yergin 1991, pp. 20–22). In 1855, at the instigation of Bissell, Silliman discovered that heating crude oil allowed the various cuts or fractions to boil off into separate marketable products. Once commercial viability was established, Bissell hired Colonel Drake in an effort to find a resource of sufficient size to make the refining process practical.

From an economic perspective, the shift to kerosene had a relatively minor economic impact and ultimately faded to insignificance as electric lighting replaced kerosene lamps. Ironically, the early waste products of refining—fuel oil and gasoline—ended up building the petroleum industry, as they began to compete with coal. Unlike the shift from whale oil to kerosene, the transition from coal to oil was highly complex and prolonged, arguably lasting over a century. The shift was complicated because it involved major changes in extraction, conversion, transportation, and end-use technologies. More succinctly, the entire infrastructure of energy production and use changed along with much of the economy.

It is useful to review what contemporaries in the nineteenth century thought about the supply and use of coal. William Stanley Jevons, an economist and Cambridge academic, became famous for his book *The Coal Question*, published in 1865.[1] Jevons reasoned that England's economic prosperity and progress was largely due to its coal, which was, of course, a depleting resource. He calculated that the country's reserves would last less than a century and that their decline would bring England's industrial revolution to a screeching halt. Critics of the peak oil thesis often cite Jevons' study, since he did not foresee that oil and natural gas would replace coal as primary energy sources and, thus, his concerns were largely unfounded. Although Jevons presented a far more sophisticated view of the world than his predecessor Thomas Malthus, his vision of the future turned out to be too narrow.

Jevons argued that: "This question concerning the duration of our present cheap supplies of coal cannot but excite deep interest and anxiety wherever or whenever it is mentioned: for a little reflection will show that coal is almost the sole necessary basis of our material power, and is that, consequently, which gives efficiency to our moral and intellectual capabilities" (Jevons, I.6). Jevons illustrated his argument by showing that the population of England and Wales had been more or less constant for centuries, but with the development of coal and the steam engine, it tripled in two generations.

Jevons did not focus solely on coal supplies; he also analyzed energy demand. Indeed, his book may be the first study of energy conservation. He presented data indicating that, from the eighteenth to nineteenth century, the energy efficiency of engines increased tenfold. However, as efficiency improved, the engines could be adapted to more and more useful work: "But such an improvement of the engine, when effected, will only accelerate anew the consumption of coal. Every branch of manufacture will receive a fresh impulse—hand labour will be still further replaced by mechanical labour" (Jevons, VII.21). This observation is known as the *Jevons Paradox*—improved energy efficiency frequently increases, rather than diminishes, consumption.

Over the decades since, the Jevons Paradox has held. In Jevons' time, England's domestic coal consumption was about 1 ton per person; in 2008, the UK per capita energy consumption was the equivalent of about 5 tons of coal. This has led some critics to conclude that energy conservation is hopeless; savings in one sector simply promotes consumption in another. Such a conclusion, however, fails to consider the underlying economics, which Jevons clearly had in mind.

The increase in per-capita energy consumption over the last century is due almost entirely to an increase in the standard of living; it is not a perversion of conservation or improved efficiency. If the British wanted to maintain their present lifestyles, including central heating, transportation on demand, mechanized tools, entertainment, and all the other trappings of modern life while still depending on nineteenth century-style coal supplies, they would be up to their eyelids in soot. It would not be feasible.

Figure 8–1 illustrates the Jevons Paradox. In this illustration, energy-efficiency improvements shift the demand schedule to the right. The shift would ensure increased consumption if energy supply could expand at constant or lower cost. On the other hand, if expanded energy supplies result in rising costs, then the precise impact is indeterminate. What matters are demand and supply price elasticities. Figure 8–1 illustrates a circumstance in which energy costs and consumption both rise. Even with higher costs, however, energy consumers are better off. This is because energy is more valuable than it was—it can facilitate an expanded list of useful activities, making life more enjoyable and less burdensome.

Jevons' most important message, however, is not his dire prediction of coal depletion or the paradox of energy conservation, but the analysis that explains how energy use is pervasive through the whole economy and is a building block—perhaps the main building block—of economic growth. Demand for motor fuels cannot exist unless consumers own cars. Likewise, consumers will not buy cars with internal combustion engines unless there is assurance of fuel supply. As Jevons explained: "It must be remembered that the progress of any branch of manufacture excites a new activity in most other branches, and leads indirectly, if not directly, to increased inroads upon our seams of coal" (Jevons, VII.7). Aside from the problem of the highly inelastic demand and supply functions discussed in chapter 2, this has strong implications for the path of transition from oil to alternatives.

Major energy transitions such as coal to oil, have been incremental and evolutionary, not revolutionary. In some sense, oil was an immediately superior fuel to coal (lower cost to transport and use and with fewer pollutants); nonetheless, it took a century to make the adjustment. The gradual shift from one energy

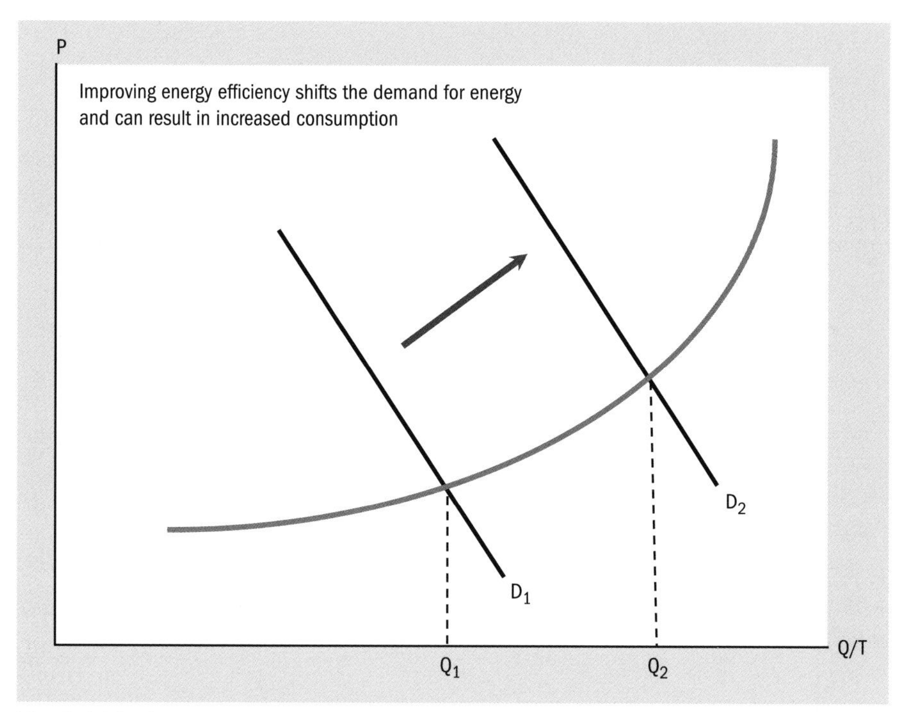

Fig. 8–1. The Jevons Paradox

source to another should not be a surprise, since a modern economy cannot function without a complex infrastructure designed around specific types of energy resources. Likewise, demand for particular energy products cannot exist without that infrastructure in place. A modern economy is a tightly interwoven package of energy production and use.

Large, complex energy systems change slowly because they require massive long-term capital investments. In contrast, the high tech industries comprising computers, cell phones, software, and related products can change quickly, with a turnover of three years or less. Despite the short shelf life, the high-tech industry exhibits many of the same characteristics as the energy system. Personal computers (PCs,) like the energy industry, depend on a series of interrelated components. Once a consumer has chosen a computer system, a legacy of files and applications accumulates, and previous choices burden the selection of new products. Likewise, an infrastructure of software and peripheral support grows up around successful computer systems. When the personal computer industry began in the late 1970s, it was a wholly new approach to office organization, a carte blanche to innovation. Thus, Apple Computer could afford to jump from the Apple II to the Macintosh technology and still stay in business. However, when Steve Jobs tried to shift customers from the Mac to the "NeXT" workstation, the jump failed. Most consumers had invested too much in their existing computer systems, software, and files. In contrast, Microsoft developed Windows at a slow pace and in manner that integrated legacy files with the new software. From a commercial point of view Windows was the big success, even though most people think the Mac or NeXT operating systems were superior products.[2] In tandem with this experience, managers in the high tech industry concluded that a wholly new technology could only succeed if it delivered a massive improvement in productivity. Typically, innovation and improvements come in a much slower and incremental manner–evolution instead of revolution.

Recently, the Japanese have invented a term-of-art for a technology too advanced or unusual to integrate into an existing system—the *Galapagos syndrome* (*New York Times* 7/20/09). It turns out that Japanese cell phones are the most advanced anywhere. Japan's phones have so many sophisticated features (from credit card uses to HD television), they are unable to find a market outside Japan, because the benefits associated with the phones require a support infrastructure enjoyed by no other country. The cart got ahead of the horse.

It is possible to imagine a revolution in the energy system something like the early years of the PC. If, for example, someone came up with the mythical engine that could run on water, there would be rapid market penetration and a complete change in fuel production, distribution and use. Barring such an unlikely event, however, oil will phase out gradually. New energy sources will emerge, but only when the pieces of supporting infrastructure are assembled.

Figure 8–2 presents an overview of motor fuel production and use and how the activity interweaves with a variety of products and services. The story begins in the middle of the chart—consumers must make a choice about how to meet their transportation needs. The choice depends on four fundamental factors: the availability and price of motor fuels, the price and types of vehicles for sale, the degree of access to roadways, and the convenience and cost of public transportation.

Behind each consumer's choice, there are thousands of supporting activities by, literally, thousands of different companies. For example, the inherent value of buying a car in New Hampshire depends, at least to some extent, on the quality of freeways in California, because the motorist must see the purchase as a universal solution to the desire for travel and transportation. Vehicle choice, and hence fuel choice, depends on a complex array of moving parts—car designers in Detroit, corn farmers in Iowa, oil production in Alaska, refineries on the Gulf Coast, highway construction in Montana, and a widespread and reliable fuels distribution system. The extraordinary drop in car sales in late 2008 and 2009 was not just a feature of the recession and lower income; it reflected uncertainty about oil prices and the dependability of auto manufacturing companies, many of which were near bankruptcy.

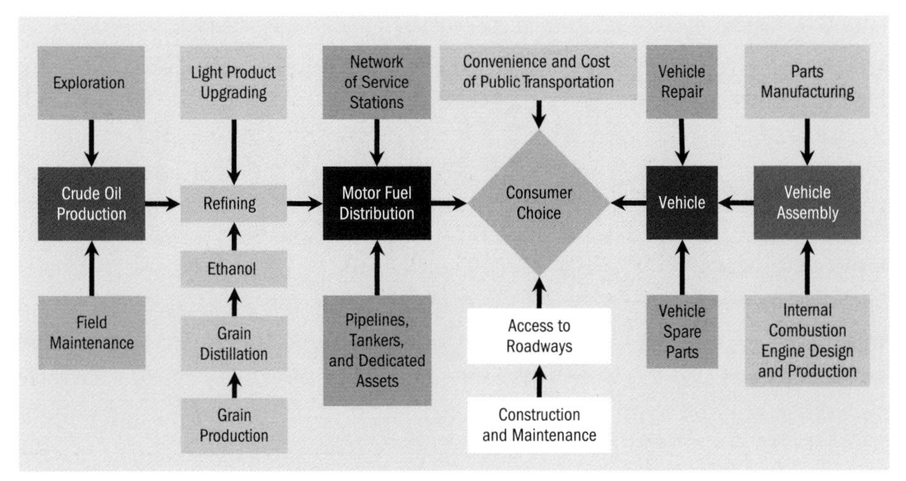

Fig. 8–2. Interwoven motor fuel supply and demand

A simple thought experiment illuminates the complexity. Assume there was no oil and gas, forcing a modern economy back to a coal-based energy system. Trains and trolleys would replace cars and trucks. Coal furnaces would replace oil and gas. The airline industry would shrink, pandering just to the rich. Car manufacturing would collapse. Service stations would disappear. Suburbs would fade and urban centers would rise. Rail would replace freeways. A much smaller number of cars would run on costly ethanol. In short, almost the entire infrastructure of the modern world would change.

Now reverse the thought process—look forward to a world with diminishing oil supplies. Most analysts simply assume that various synthetic oils will replace conventional petroleum products. That is not, however, what happened when coal replaced oil. Coal use today is more or less the same as it was a century ago. Instead, the availability of oil provoked a set of technological opportunities to use energy in new ways. For example, no one would have conceived a jet engine without oil. Synthetic fuel production is inherently inefficient (in both energy losses and environmental side effects). Consequently, these options are unlikely to play a big role in the forthcoming energy transition. Instead, new types of energy sources will be married to new types of use technologies in an evolving energy system.

Table 8–1 provides a snapshot of key elements of the oil and gas industry in the U.S. in 2002. The table excludes a wide variety of peripheral activities, everything from plastics to asphalt manufacturing. Nonetheless, the table provides a concrete example of the extent to which the oil industry percolates through a modern economy. Even though their numbers are declining, gasoline service stations alone employ 6.3% of the total workforce in retail trade.

Table 8–1. Highlights of the U.S. oil and gas industry's 2002 economic footprint

Sector Activity	Crude Petroleum and Natural Gas Extraction	Support Activities for Oil and Gas Operations	Natural Gas Distribution	Oil and Gas Pipeline Construction	Petroleum Refineries	Gasoline Stations
2002 NAICS Sector	211111	213112	2212	237120	324110	447
Number of Establishments	7,178	6,297	2,376	1,403	199	121,446
Sales, Shipments, Receipts	$85,906,216	$11,374,939	$66,515,186	N/A	$194,761,863	$249,141,412
Annual Payroll	$5,036,621	$3,887,423	$5,369,818	$3,984,827	$4,395,947	$13,700,950
Number of Employees	94,886	111,481	85,447	93,176	62,540	926,792

Source: US Census 2002.

Fortunately, there is still a lot of conventional oil and gas left in the globe and no need to hit the delete button. Moreover, the economic cost of such a high-speed transition to other energy sources would be astronomical. Barring an unforeseen catastrophe, oil will gradually phase out just as did the market for whale oil; indeed, the transition to other energy sources has already begun. In the 1960s, oil was so cheap it pushed out nearly every other fuel. Coal mines in Europe and Japan were shut down as electric utilities and industrial customers shifted to fuel oil. The price rise of the 1970s reversed the trend. As noted earlier, about three-quarters of oil use is now in the transportation sector, where it has a distinct advantage. As new technologies and innovations emerge, however, oil's last stronghold will succumb, raising an inevitable question: What are the emerging technologies for transportation and which are likely to prevail?

Comparing the Alternatives

In what must have been a frustrating meeting, President Truman is said to have exclaimed, that from now on he would hire only one-armed economists—he was far too tired of hearing, "on the other hand." It is a good thing that centipedes do not study economics, or the President would have thrown a real fit. A comparison of alternative automobile technologies looks very much like the life's work of a centipede. The comparison is not simple since it involves a wide variety of objectives, and no one technology is superior in all categories. Put plainly, there are trade-offs, and no choice is perfect—a classic "on the other hand." Analysis is made all the more difficult because most of the technical studies of the principal options are advocacy arguments in disguise, and typically focus on the superior features of the favored technology, while ignoring the drawbacks. It is tough to see through the clamor of special interests and crackpot ideas. Here the objective is different: What are the positives and negatives of each option and what is the marketplace likely to choose? Table 8–2 surveys the most likely options to replace conventional oil in transportation, with respect to the various characteristics and tradeoffs.

Table 8–2. Thumbnail sketch of alternative energy for transportation

	Synthetic Fuels (Sands and Shales)	Compressed Natural Gas	Hydrogen Fuel Cells	Gasoline Hybrids	Electric Battery (Li-Ion)
Resource	Huge, but great quality variation	More abundant than crude oil	Multi-source flexibility	PHEV has multi-source flexibility	Multi-source flexibility
Cost	Much higher than conventional oil	Dependent on low-priced natural gas	Not yet competitive	Near term, most economic	Cost dependent on battery
Infrastructure	Processing and Distribution in place	Limited, mature technology	Limited, difficult technology	In place	Homes, parking lots, RV parks
Technological Status	Oil Sands only proven technology	Mature Industry	R&D	Presently penetrating markets	Commericalization in Process
On Board Storage Vessel	Conventional tank shaped to car	Heavy, constrained to cylinder	Heavy, shape and structural issues	Conventional tank shaped to car	Battery heavy, but shaped to car
Energy Density on Vehicle (Volume)	34.2 MJ/L	9.3 MJ/L	5.6 MJ/L	34.2 MJ/L	1.0 MJ/L
Energy Density on Vehicle (Weight)	46.7 MJ/kg	53.6 ML/kg	141.9 MJ/kg	46.7 MJ/kg	0.6 MJ/kg
Well-to-Wheel Efficiency	0.41 km/MJ	.32 km/MJ	0.35 km/MJ	0.64 km/MJ	1.14 km/MJ
CO_2 Emissions	46.8 g/km	45.0 g/km	0 - 41.1 g/km	31.2 g/km	0 - 12.6 g/km
Environmental Impacts	Major side effects in production & use	Least polluting hydrocarbon	Pollution free use	Lower than conventional cars	Pollution free use
Reliability/Safety	Comparable to present technology	Trade-off between weight and safety	Risks associated with Hydrogen	Comparable to present technology	Battery fires
Security	Domestic resources are secure	Mainly domestic production	Fuel switching improves security	Fuel switching improves security	Fuel switching improves security

Sources: Author, Tesla, BP, DOE.

The first item considered is the primary energy source itself: Is the resource large enough to support multi-decade periods of development, use, and eventual obsolescence? Recall Henry Ford's concern about basing the viability of his automobiles on petroleum, which could become scarce at any time. Now consider the present system. Vintage cars abound—the oldest ones appear mainly in parades and rallies, but at any one time, the active fleet, used on farms and highways, varies in age from thirty minutes to thirty-plus years. The vast bulk of the vehicle stock will have to have infrastructure support through their expected lifetime, which is certain to be longer than a decade. Even if a vastly superior alternative to conventional fuels were discovered tomorrow, the capital stock of existing vehicles and supporting infrastructure would take a generation to phase out.

A second issue is the production and processing cost of the alternative resource. Many resources can substitute for conventional crude oil and petroleum products. For example, oil shales, oil sands, and heavy oils are abundant, and the resource could last for centuries. With the exception of heavy oils and oil sands, however, only a few small-scale, and largely experimental projects, are in place because costs are too high.

As already pointed out, infrastructure issues are a major hurdle for hydrogen fuel cell vehicles and, to a lesser extent, natural gas and all-electric vehicles. On the other hand, hybrid vehicles, including the plug-in hybrid electric vehicle (PHEV,) mainly use the existing infrastructure, so any transition is streamlined.[3]

All the options listed in table 8–2 are based on proven technologies from various R&D projects. However, the prospects for improvement—better performance at lower cost—vary significantly. The variation depends mainly on the extent to which the technology can take advantage of mass production techniques that will ultimately reduce cost. Right now, most of the experimental cars are hand-built and consequently quite expensive. It is in such analysis that biases inevitably creep in; if the technology is not economic now, then a simple assumption can make it seem to be so.

Energy has to be stored to be portable and that is a major constraint for some of the options in table 8–2. There are three planning elements—total volume, weight, and the flexibility of the vessel's shape. Conventional oil is the clear winner in this category. Neither gasoline nor diesel fuels require heavy tanks; the tanks are shaped to the car's undercarriage and do not have to be pressurized. In contrast, most of the alternative energy sources have complex and costly storage vessels.

There are two useful measures of a fuel's "energy density," volume and weight, and both are relevant to transportation where space and burden are at a premium. Table 8–2 measures volumetric energy density in mega joules (MJ) per liter (L). A quick glance at the table demonstrates the clear advantages of liquid fuels. In comparison to gasoline, natural gas requires three times the storage volume to generate an equivalent amount of energy, while hydrogen requires five times the volume.

Liquid fuels do not have an advantage over gases when it comes to weight. Hydrogen has three times the thermal energy of gasoline or diesel for the same number of kilograms. That should not be a surprise since hydrogen is the lightest element in the periodic table. Natural gas has slightly more energy per kilogram than do petroleum products. The need for heavier and less flexible storage vessels, however, offsets the weight advantage of these fuels. Table 8–2 compares the various technological options in MJ per kilogram (kg).

Simple measurements like MJ/kl or MJ/kg are only a small part of the energy efficiency calculation. There is a long chain of production, conversion, and distribution of motor fuels. From an energy perspective (and by implication, the perspective on environmental side effects) what matters is the "well to wheels" efficiency. Starting out with the bulk form of crude oil (or an alternative energy source) how much of the inherent energy actually turns the wheels and drives the car? This is where electricity has a clear advantage over liquid or gaseous fuels. An electric motor is approximately four times more efficient than an internal combustion engine, which dissipates most of its energy as heat.

The well-to-wheel efficiency figures in table 8–2, are in kilometers (km) per MJ. These figures are from Eberhard and Tarpenning (2006, p. 3) of Tesla, an electric car manufacturer, and describe a Honda Civic in each category except the electric battery, which describes the Tesla sports roadster. The synthetic fuel efficiency figure in the table assumes the efficiency of conventional crude oil and petroleum products less 20%, reflecting the energy penalty associated with producing synthetic oils.[4] The figures, as listed, suggest electric vehicles are vastly superior. Some qualifications, however, are required. The Honda Civic seats four people and has room for luggage. The Tesla seats only two people and much of the trunk space holds the battery pack. The Tesla engineers also assumed that electricity generation would be from the most efficient gas generators. Taking account of these assumptions, a more realistic calculation would put electric vehicles on a par with hybrids.

It should not be a surprise that the volume of CO_2 emissions varies proportionally to the overall well-to-wheel efficiency of the vehicle. The CO_2 figures in table 5–2 were also calculated by Eberhard and Tarpenning (2006, p. 4). CO_2 emissions of synthetic fuels in table 5–2 are 20% higher than conventional oil, to reflect conversion activities. Electric vehicles (EV's) are, of course, superior, since there are no emissions while driving. An EV's CO_2 emissions are less than half of the nearest rivals, which are hybrid cars. This assumes emissions are from gas-fired electricity, which in most cases is the marginal source of power generation.[5] With this type of study, it is important to keep key assumptions in mind. Coal-fired electricity would have a much higher emission of CO_2 just as nuclear, hydro, solar, and wind would have far less. The key question to ask going forward is the incremental quantity of CO_2 and other pollutants that arise from shifting vehicle types. In most cases, natural-gas-fired generation (rather than coal or nuclear) fills incremental electricity demand, although it may not always be an efficient plant.

The environmental impact of producing and using the resource is of increasing importance, and the full life cycle of the activity should be the measure. Hydrogen fuel cell vehicles have zero pollution in use. Nonetheless, the energy has to come from somewhere and the most likely source is natural gas, which is converted to hydrogen through a reforming process rather than electrolysis. This process actually requires more energy and produces more CO_2 than using efficient gas-fired generation.

Reliability and safety are also big concerns for the motoring public. Processing synthetic fuels into conventional petroleum products has no consequence to the vehicles that will use the fuel; as a fuel, it is indistinguishable from conventional supplies. Likewise the hybrid technology is now well developed. It is also relatively simple to convert a gasoline-powered car to natural gas—the main problems are the pressurized tanks. At the other extreme, hydrogen fuel cell vehicles represent a major shift in technology that has safety and reliability implications. Lithium ion batteries also have safety implications. Improperly designed, they have caused laptop fires, and, in a vehicle, they will have to be spaced in a manner that prevents overheating.

A final consideration of future fuel choice is in the last line of table 8–2—energy security. Fundamentally, there are two criteria: 1) Does the primary resource have to be imported, and 2) Are there multiple options for producing the primary energy resource and is fuel switching practical for the consumer?

Synthetic Fuels

Fuels made from corn, sugar, oil sands, and oil shales are energy hogs because it takes a lot of energy to produce and convert the raw material into usable products. The conversion is necessary because these raw materials are too bulky and too laden with contaminants for direct use by engines. In contrast, the conversion of crude oil into useable products is relatively trivial. Were it not for the environmental side effects of synthetic fuel conversion, these options could provide adequate energy for decades, if not centuries. The resource base is truly impressive. However, the impact on global warming, air quality, and land use seriously constrains development.

After several decades of experimentation, in the early twentieth century, it became clear that liquid fuels would dominate automobile technology. Liquid fuels had both superior energy density and ease of handling, making them ideal for cars and trucks. This leads to a key question, then and now. Are alternative forms of energy delivery and storage realistic for small vehicle transportation? If not, then synthetic fuels technologies that reduce the energy penalty and constrain the environmental side effects will have to be developed.

Although synthetic fuels are costly to produce, there is an obvious advantage, since there is no need to redo most of the supporting infrastructure. With relatively minor adjustments, diesel engines can run on biodiesel. Likewise, there are now "flex-fuel" cars that can run on gasoline, ethanol, or some combination. Volvo has even designed a car that can run on five different fuels including methane and hythane (a blend of hydrogen and methane).

Due to reliance on domestic resources, synthetic fuels are secure—they were the centerpiece of early attempts by the Nixon and Carter administrations to reduce oil imports. The security, however, comes at a price. To maintain U.S.

domestic production, the federal government must offer substantial subsidies for both corn ethanol and synthetic gas from coal. Sugar-based ethanol from Brazil and other tropical countries costs less but is obviously less secure. As noted, producing fuel from grains and sugars drives up food prices. So far, it is not economical to produce ethanol from cellulous biomass or from algae.

Compressed Natural Gas

Canada, Argentina, and other locations with abundant gas reserves use natural gas as a motor fuel. Gas also powers buses and fleet vehicles in the U.S. and a variety of countries. T. Boone Pickens, among others, put forth a plan to shift to compressed natural gas in motor vehicles as a means to reduce oil imports. Pickens sees the policy as a bridge to hydrogen or electric cars. The Pickens' plan, however, presumes a massive expansion of domestic natural gas production, which is probably not realistic. In 2008, U.S. oil imports were 641 million tons, while total gas production was 533 million tons of oil-equivalent. Thus, a 50% reduction in oil imports would require something like a 60% rise in gas production. LNG imports could not be relied on either; in 2008, *total* world exports were 204 million tons of oil equivalent, about 38% of U.S. oil imports (BP 2009).

The gains from the Pickens' plan are unclear, at best. Shifting dependence from imported oil to imported LNG would not appreciably improve energy security and could provoke a successful gas cartel to match OPEC. Although there has been an increase in domestic gas supply, particularly from unconventional sources, almost no one believes that supply could ramp up quickly enough to make a major dent in oil imports. Moreover, even Pickens admits that the proposal is only a bridge to other more sustainable energy supplies. Thus, shifting large amounts of the energy delivery infrastructure to accommodate compressed natural gas vehicles has a good chance of being a blind alley.

As already noted, storage is an important constraint for using compressed natural gas (CNG) on a vehicle. The tanks used in taxis and other fleet vehicles do not hold a great amount of fuel. The Honda CNG is typical of this type of vehicle and has a range of 170 miles (www.fueleconomy.gov). In order to get a longer driving range, there has to be either more gas cylinders or greater compression with heavier and stronger tanks. This reduces both the efficiency of the vehicle and the amount of luggage space. On the other hand, some pickups and other converted vehicles where space is not a constraint have simply added tanks, increasing the driving range to 600 miles or more. Such conversions have an additional advantage in that the vehicles can switch back to gasoline if CNG is not available. According to Natural Gas Vehicles America (NGVA), there are 1,100 refueling sites in the U.S. to serve over 120,000 NGVs in the U.S. Worldwide there are now 8.7 million vehicles.

There are two advantages to CNG vehicles. Since the oil and gas price collapse of the summer of 2008, North American natural gas prices have been substantially cheaper per MJ than oil products. Given that well-to-wheel energy efficiency for gas and oil vehicles are about the same, it is likely to be much cheaper to run CNG vehicles over the next decade. This explains the increasing popularity of CNG for taxis and other fleets, where refueling is not a constraint. The other advantage is security. In sharp contrast to oil-import dependence, the U.S. imports natural gas and LNG more as a convenience than a necessity, with net imports in 2008 representing only 12.8% of total consumption. In the event of an oil supply disruption similar to those during the Arab oil embargo and Iranian Revolution, CNG vehicle owners will be able to run uninhibited, while other Americans would sit in filling station queues.

Although compressed natural gas vehicles are not likely to reduce oil imports, they could be quite helpful in moderating high oil prices. In other words, the Pickens' plan in moderation is good idea. Events in 2008 and 2009 once again demonstrated that small changes in global oil demand or supply have a huge impact on oil prices. Multi-fuel vehicles, such as the one designed by Volvo, or converted gasoline cars and trucks, present an interesting option. Infrastructure that allows consumers to switch to different types of fuel supplies could help moderate energy price swings.

Hydrogen Fuel Cells

NASA installed fuel cells rather than batteries for its space flights because it was far more important to keep weight low than to contain cost. But NASA did not have to worry about a supporting infrastructure either, nor could it rely on one. Once in space, everything has to be self-contained; there are no service stations on the moon. Although the hydrogen fuel cell technology is fascinating, its adoption would require significant shifts in vehicle design and infrastructure support, more so than any other option.

As originally envisioned—as part of a nuclear fission-fusion complex—there would be no resource constraints on hydrogen production. Likewise, hydrogen produced from solar, wind, or hydro sources has minimum side effects on air quality or the production of greenhouse gases. (Increased water vapor could have an impact on global warming similar to increased CO_2.) As pointed out earlier, however, a large number of MJ are lost in the conversion of electricity to hydrogen. At present, the distributed steam methane reformation (DSMR) technology using natural gas as a feedstock is the more practical and cheaper option for the near future (NAS 2008, p. 5). Longer term, the National Academy of Sciences study authors believe coal or biomass gasification would source the hydrogen. Coal gasification would include carbon capture and sequestration.

Such a complex supply chain is difficult to envision—coal to gas and then hydrogen to electricity. No matter how good the technology, multiple conversions are going to result in poor energy efficiency. Producing synthetic gas from coal normally has a 20 to 40% energy loss. Fuel cell energy conversion efficiency—from hydrogen to electricity—can be as high as 50%. However, according to Eberhard and Tarpenning (2006), the actual efficiency in the Honda FCX has an implied fuel cell efficiency of around 31%. All of these losses compound, and by the time power gets to a motor, only a fraction of the energy is left.

The energy losses from coal to hydrogen to electricity are far greater than the loss of producing crude oil and refining it into products suitable for transportation fuel. However, better engine design compensates for much of the deficit. Electric motors drive the wheels directly, avoiding the inefficiencies of the internal combustion engine. In the end, the well-to-wheel efficiency of the hydrogen-fuel-cell car ends up about the same as conventional gasoline cars or the natural gas alternative if natural gas produces the hydrogen. But the technology begs a question: if electricity is the most energy-efficient way to power a vehicle, why not skip the middleman and find a way to use it directly?

Fairly or unfairly, there is public concern about using hydrogen as an energy source. This stems from the spectacular fire and crash of the German rigid airship, the *Hindenburg*, in 1937. Popular anxieties aside, there are a number of technical issues regarding the use of hydrogen that will have to be solved. Over time, pipelines and other vessels that contain hydrogen can become brittle as the tiny molecule embeds in the steel. Interestingly, hydrogen burns without a visual flame. This led NASA to conceive the "broom" test. That is, to find out if the hydrogen was burning, experimenters placed a straw broom in the path of the flame, a low-tech solution for a high-tech organization.

Hybrids

El Pueblo de Nuestra Señora la Reina de Los Angeles de Porciúncula, better known as Los Angeles, was a land of palm trees and movie stars until World War II. Then, in 1943, it experienced its first powerful attack of smoke and fog, abridged quickly to "smog." Rapid post-war population growth and incomprehensible freeway expansion compounded the problem and by the 1960s, the city was frequently invisible beneath a brown smudge, launching a powerful constituency for clean air. Federal environmental rules on motor exhaust eventually reversed the trend, but ever since, California has sought to lead the nation in clean air standards, mandating its own gasoline blends, fuel-burning restrictions, and even car design regulations.

In 1988, the legislature passed and Governor Deukmejian signed the California Clean Air Act, which set the regulatory framework for the coming years. Following the mandate of the act, the California Air Resources Board (CARB) laid out a plan aimed at moving the state's commuters to zero-emission-vehicles (ZEV). The motivation was, of course, to clean the air, not to make cars more efficient.

The 1990 CARB plan was bold. It mandated minimum percentages of ZEV sales: 2% in 1998, 5% in 2001, and 10% in 2003 (PBS NOW 2005). The primary options were battery electricity vehicles or hydrogen fuel cells. Car manufactures were less than thrilled with the mandate, immediately recognizing the disadvantages of producing cars for a single state, no matter how large. Nonetheless, a pilot program got underway, with four car companies participating.

GM introduced its electric car, the EV1, in 1997 (fig. 8–3). The car was a two-seater with batteries taking up much of the remaining space. Initially, the car had a limited driving range—less than 100 miles—and ultimately was withdrawn from the market. In 2007, *Time* magazine listed the EV1 as one of the 50 worst cars of all time, but rather than damning with faint praise, it praised the damned: "The EV1 was a marvel of engineering, absolutely the best electric vehicle anyone had ever seen. Built by GM to comply with California's zero-emissions-vehicle mandate, the EV1 was quick, fun, and reliable." One car ended up in the Smithsonian Museum. In describing the car, the museum noted that the EV1 took seven years to design and build and it was indeed a 'marvel of engineering': "The automobile combined lightweight materials, aerodynamic design, a system that recharges the batteries during braking, and sophisticated computer-controlled propulsion to create an electric vehicle."

Fig. 8–3. 1997 Generation 1 EV1 electric car

GM's primary problem was its battery. If electric cars were to fulfill their promise, they would require a robust lightweight battery. Initially, the EV1 had a traditional lead-acid battery design—this is why there was only room for two adults. In 1994, GM purchased the patent rights to a new type of battery design, the nickel metal hydrite battery (NiMH), and set a joint company to manufacture car batteries. Similarly, Japanese car manufactures formed a consortium with Panasonic, which developed the EV-95 NiMH battery. GM was first with the electric car, but Honda and Toyota were the first with good batteries, marketing the Honda EV Plus and an all-electric version of the Toyota RAV 4 in California to a limited number of buyers. GM soon followed with NiMH batteries for the EV1, which reduced weight and extended the range to 120 miles.

Despite the enthusiasm of EV1 drivers and those that were unable to buy or lease them, the California experiment ended as just that, an experiment. The problem, according to GM, was the car's excessive cost and the limitations imposed by the battery. GM recalled the EV1 in 2003, crushing most of the cars in a symbol not lost on environmentalists. Many EV1 leasers actually joined in a "funeral procession" when the model was recalled (*Smithsonian* 2008). (GM was concerned that without qualified battery maintenance, there would be safety hazards.) Honda and Toyota left their electric cars in place, but did not manufacture any more. The CARB retreated from its aggressive ZEV targets and ultimately, in 2008, shifted focus to hybrid and plug-in hybrid cars.

To this day, controversy engulfs the California ZEV initiative. Critics claim that the experiment was set up to fail; the potent interests of big oil and auto manufacturing shut it down in an effort to preserve their markets. In 2000, GM did sell its interest in the NiMH battery manufacturing company to ChevronTexaco, and the co-inventor of NiMH batteries, Stan Ovshinsky supports the suspicion of conspiracy. In a 2008 interview, he claimed: "I think we at ECD [the company founded by Ovshinsky] made a mistake of having a joint venture with an oil company, frankly speaking. And I think it is not a good idea to go into business with somebody whose strategies would put you out of business, rather than building the business."

These suspicions, however, overlook economic fundamentals. The California ZEV experiment did not flounder because of a conspiracy, poor engineering, lack of enthusiasm, or poor technology; it failed due to awful timing. As Californians were tooling around their urban centers in slick and peppy EV1s, U.S. gasoline retail prices were scraping bottom, averaging $1.07 per gallon in 1998 and $1.18 in 1999. It just did not make sense to build cars for a mass audience with limited passenger space and multi-thousand dollar batteries. American consumers were doing exactly what market signals were telling them to do, and U.S. auto companies were simply following their lead.

It is often the case in science that a failed experiment adds more to the body of knowledge than a success. The experimental technology developed and proven for California's electric cars, particularly regenerative braking and the NiMH battery, helped open the door for a whole-new type of efficient car—the hybrid. Ironically, however, it was Japanese companies who profited from the California ZEV experiment, rather than American car companies.

In 1993, the Clinton Administration founded the Partnership for the Next Generation of Vehicles (PNGV). The object of the partnership was to coordinate the development of advanced automotive technologies among car companies in order to achieve the greatest feasible advance. Participants included U.S. auto manufactures, but excluded foreign manufactures, even if they had plants in the U.S. Miffed, Toyota decided to undertake its own research in secret (Prius History 2009). The result, four years later, was a midsize hybrid car, the Prius. Until thoroughly tested, Toyota marketed the Prius only in Japan; worldwide distribution began in 2000. The Prius was an immediate success, particularly in the U.S.

U.S. engineers were hardly ignorant of the hybrid concept. In 1995, at the request of Congress, the Office of Technology Assessment (OTA) completed a study of advanced automobile technologies that reviewed all the usual suspects: natural gas, electric, and hybrids. It concluded that: "advanced vehicles are likely to cost substantially more than their conventional counterparts, and the savings resulting from their lower fuel consumption will not offset their higher purchase prices" (OTA 1995). That is still, more or less, the case. The 2010 Prius starting price is around $22,000 and averages about 50 miles per gallon (mpg). The most efficient conventional American car in the Prius class, the Saturn Astra starts at $16,495 and averages 28 mpg.[6] At $3 per gallon, 15,000 miles per year, and a 6% discount rate, it would take ten years for a consumer to recoup the extra $5,505 in car price. Another comparable midsize car is the Nissan Versa, and it has an even lower starting price, suggesting that the Prius sells for a premium for reasons other than just fuel efficiency.

Motorists shopping for a new car find a vast array of makes and models. The Fuel Economy Guide from the DOE and EPA classifies 14 different classes of vehicles, from two-seaters to standard pickups. The 2009 report calculated fuel efficiency for 188 different vehicles. There is a great deal of space for individual preferences as well as efficiency considerations. Some buyers are willing to pay more for cars that are cleaner and have less impact on the environment; others are "early adopters" who want to be the first to select a high-tech option. The sheer diversity of choice makes it difficult to pinpoint the precise willingness of consumers to pay for more efficient car designs, but clearly some are.

Electric Vehicles

Batteries for hand tools and laptop computers have steadily improved; they now last longer, are more reliable, and have a greater energy density than ever before. Cost has gone down too, typically at 6 to 8% per year. The building block for these devices is the lithium-ion (Li-Ion) battery. The promise of this new technology, combined with the oil price spike of 2008, sparked a near stampede among car manufacturers. Table 8–3 lists the variety of electric and plug-in hybrid vehicles now available or planned for release in the next few years.

After decades of R&D, electric vehicles are finally ready to enter the mass market. The principal drawbacks of the design remain—the vehicles have limited range, recharging takes time, and the cars cost more. Nonetheless, these constraints are slowly eroding, as improved technology and the benefits of mass production promise better and cheaper vehicles.

Table 8–3. Sampling of electric vehicles in production or planned

Maker	Type	Car	Battery Pack	Battery Range	Date to Market	Location	Estimated Price
Nissan-Renault SA	EV	LEAF	Li-ion	100	2010	Japan, U.S.	$20,000 Lease Battery
Mitsubshi	EV	I MiEV	Li-ion	100	2009	Japan	$22,000
Tesla	EV	Model-S	Li-ion	300	2011	US	$49,000
BYD	EV	E 6	Fe	248	2009	China, US	NA
Subaru	EV	STELLA	Li-Ion	55	2009	Japan	$29,000
BMW	EV	Mini E	Li-Ion	156	2010	Germany, US	Lease $820 month
Ford	EV	Th!nk City	Mes-dea Li-Ion	112	2008	Norway, Denmark	$19,800
Daimlier AG	EV	Smart ED	Sodium-Nickel	71	2010	UK, Germany	$20,000
Axiam Mega	EV	City	Lead Acid	60	2008	UK, France	$20,750
REVA	EV	G-Wiz	Li-ion	75	2008	India, Europe	$26,000
Tesla	EV	Roadster	Li-ion	244	2008	US	$100,000
Toyota	PHEV	Prius Plug-in	Li-Ion	7	2011	U.S.	$48,000
BYD	PHEV	F3DM	Fe	60	2008	China	$16,000
GM	PHEV	Volt	Li-Ion	40	2010	U.S.	$43,000

Other Notes: Japan subsidy $15,000; US subsidy $7,500; UK Subsidy, $8,250; China Subsidy $7,500.

Like the four points of the compass, there are four distinct paths of electric vehicle design. Some companies, such as GM, are focusing on expanding the hybrid car concept—extending the range of battery-driven miles, while retaining the capacity to switch to conventional fuels. Other companies are concentrating on producing small urban cars, like the Ford Th!nk City, designed for limited daily use at slower speeds. There is an emerging market for high-performance EVs, with the start-up Tesla firm leading the way. Most companies are concentrating on creating a family-oriented vehicle for commuter driving; ultimately, longer distance trips at conventional speeds, comparable to gasoline and diesel vehicles. Figure 8–4 illustrates the four categories of vehicles.

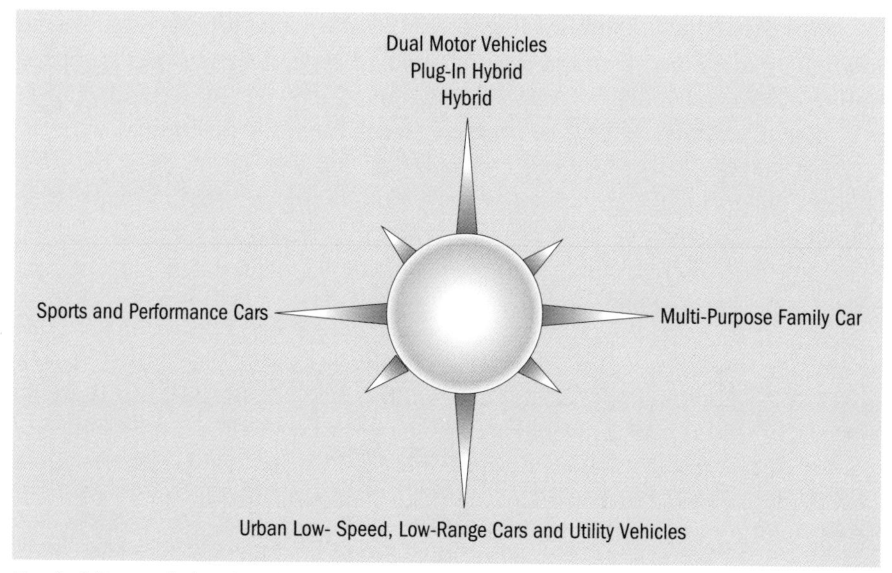

Fig. 8–4. Types of electric vehicles

The plug-in hybrid electric vehicle is the next step in the hybrid saga. It, like the hybrid, combines electric and conventional fuel propulsion. There are, however, several distinct differences. The PHEV will be able to run on the battery alone before it switches to gasoline. In some cases, the car will have an electric generator to charge the battery and drive the wheels' electric motors rather than a conventional gasoline engine. The car's battery charges at home from a conventional electric plug or an upgraded 240-volt connection, hence the name "plug-in."

The most widely publicized and anticipated of these new vehicles is the Chevrolet Volt, due out in 2010. GM has poured enormous effort and investment into the Volt and the transition to electric technologies. In June 2009, GM's Vice President for Global Engineering, Jim Queen, stated on the opening of GM's research and test facility, "GM's comprehensive battery strategy, which includes ramping up in-house responsibility for advanced battery technology... will be the keys to sustainable transportation." GM's new test facility is three times larger and is able to test a wide variety of battery designs for durability and longevity.

The concept behind the PHEV is straightforward. Most car journeys are less than 40 miles. If the commuter plugs in at night, driving to work or school and back, the onboard generator may never turn on. GM calculates that battery-driven propulsion will account for 70% of the car's use. On the other hand, if a longer journey is necessary conventional fuels will be there as backup. Because the PHEV runs in the two modes, the lack of a robust recharging infrastructure is not a constraint.

There are, however, interesting alternatives to the Volt. Toyota plans to extend the battery range of the Prius to 7 miles and add a plug-in capability. This is not as ambitious as the Volt, but many commuter trips are for short distances. The Chinese battery and car company, BYD, has had a plug-in hybrid car, the F3DM, on the market since late 2008. This car has a battery-only range of 60 miles and sells for less than $20,000—recent reports put it at around $16,000. Apparently, BYD had to drop the price because sales were slow, even though there is a subsidy of around $7,500 from the Chinese government.

The Tesla car company in California introduced its roadster to great fanfare in late 2007. Hollywood celebrities jumped quickly to be first on their block to go green. (The car, by the way, is named for Nikola Tesla, one the early pioneers in the commercial development of electricity.) In many respects, the car is truly innovative. Rather than design and construct batteries from the ground up, the company bundled together off-the-shelf laptop batteries. The key design characteristics were the arrangement and organization of the batteries in a pack to prevent overheating or the problems with one battery contaminating all the others. The most extraordinary feature of the Tesla is its style and performance. It has better acceleration and handling than many similarly priced gasoline sports cars and it looks good. Thus, its price tag—around $100,000—has not so far inhibited sales.

In stark contrast to the luxurious Tesla, electric bicycles and scooters are increasingly popular, particularly in Asia, selling in the millions. The bicycle, of course, dominated transportation in China for several decades before the recent wave of economic growth. Most of these new vehicles use lead-acid batteries. However, as the market grows it will create demand for more efficient batteries, and that will help create the economies of scale and technological evolution necessary to lower cost.

There are also a range of small-scale EVs already in production with sales in the U.S., Japan, and Europe. In the U.S., they are classified as "neighborhood electric vehicles" and in the UK, as "quadracycles." Everyone is familiar with golf carts, and these are similar. For example, the GEM, from a company owned by Chrysler, has a range of 30 to 40 miles and a top speed of 25 mph. The GEM comes in a variety of designs, including one that has a flatbed capable of handling over 1,000 pounds of cargo. These vehicles are legal for neighborhood streets and industry campuses, but not for highway driving.

In the last few years, there have been a number of scaled-up, but still small, electric vehicles entering the market for urban and highway driving. The two most prominent in the UK are the Axiam Mega City and the REVA G-Wiz, the former from France, and the latter from an Indian car company. The G-Wiz is interesting because it sells with either a bank of lead acid batteries or lithium-ion batteries. The difference in price is about $12,000. Two well-known brands are poised to enter and effectively upgrade this market, the Daimler Smart ED and the BMW Mini-E.

Designers for the all-electric Mitsubishi I MiEV, Nissan LEAF, and Tesla Model S aim at the broader market, true multipurpose vehicles. There are many more planned from a variety of companies, but these are the most advanced and technologically sound. The I MiEV and LEAF are due for general sales in 2010, but are already in test markets. The Model S is credible, because of the success with the Tesla Roadster. All these vehicles use versions of lithium-ion batteries. Nissan plans to sell the LEAF as a stand-alone vehicle and lease the battery bank separately. This allows the company to keep the price of the LEAF comparable to conventional fuel vehicles; it also reduces risk for the buyer. Obviously, however, it increases monthly costs. For these vehicles, battery quality and recharging facilities are big issues. But before turning to a discussion of battery and recharging technologies, some discussion of the Chinese efforts to build an EV is worthwhile.

As China modernizes its economy, its leaders seek to shift emphasis from clothing and other low-tech manufacturing to far more sophisticated industries—aircraft, automobiles, and so on. China's biggest advantage is its huge internal market; nonetheless, it is hard to compete with established manufacturing companies in developed countries. Thus, the paradigm shift to alternative-fuel vehicles presents a unique opportunity to enter a difficult, but potentially lucrative market. BYD is one of the world's largest battery manufacturers and has been making conventional cars for nearly a decade, so electric vehicles are a natural extension of its business. BYD introduced the E6 at the 2009 Detroit Auto Show. The company also gained attention when Warren Buffet purchased stock. The E6 has a Fe (Ferrous) battery, also known as lithium-ion-iron battery, which is cheaper than batteries containing cobalt or other higher cost materials. This, along with lower Chinese manufacturing costs, allows BYD to market its planned E6 at prices half that anticipated for GM's Volt.

The key to successful electric vehicle design is the battery and it has been a roadblock for over a century. To succeed, the battery has to be as lightweight and as small as possible, not prone to overheating, sturdy enough to survive all sorts of potholes, rechargeable more than a thousand times with minimum loss of capacity, quick-charging, and able to provide surge power for acceleration. This is a tall order, and it is why the major automobile manufactures insist on extensive trials before taking an EV to market.

EVs seem to have a simple and straightforward technology, but it is more complicated than commonly thought. A modern car has all sorts of peripheral features that consumers will expect in new vehicles. Simply adding air conditioning cuts down an EV's range. Likewise, EVs will have to have heating systems for winter driving. Engineers have discovered that they achieve efficiency only through complete redesign.

With any new technology, there remains significant uncertainty about cost, and EV battery packs are no exception. GM's initial estimates for the Volt battery were far too low, and estimates are now around $12,000 for 14 kilowatt-hours (kWh). (This partly explains why the Volt's price has escalated from the mid-20s to more than $40 thousand.) However, the industry has not settled on a standard design, and initial production levels do not yet take full advantage of scale economies. The *Report of the CARB Independent Expert Panel 2007* provides an objective analysis of likely battery costs as production ramps up. The report concluded that batteries for virtually all prototype plug-in hybrids and electric vehicles had a net present value when compared to cars that depend on buying retail gasoline, including taxes. Their analysis used a robust definition of the battery characteristics (ten-year life at 80% capacity or more), but assumed high gasoline prices. The report included an extended analysis of manufacturing costs, with much of the data provided by the industry. Briefly, they concluded that battery costs would range from $1,650 for hybrids to $13,680 for full-featured electric cars suitable for the American market. The costs varied based on battery capacity, from 2 kWh to 40 kWh, and rate of production from 20,000 to 100,000 units per year (Kalhammer 2007, p. 47).

For an all-electric vehicle, battery recharging is a major issue. It only takes a few minutes to buy fuel, but recharging a battery can take all night—and woe to the commuter that fails to plug in. Standard electrical plugs in the U.S. and Japan at 110 V are too slow. The industry is in the process of adopting a standard 220 V 70 A connector for the new vehicles. This cuts charging time in half, but will require the consumer to install special wiring and adaptors. Europe will have a different standard at 400 V.

A number of competing high-speed recharging protocols are available. The idea is to provide charging stations like gasoline service stations, but likely sited at shopping centers and parking lots. The system developed by TEPCO and Nissan for the LEAF allows an 80% charge in about thirty minutes. However, the industry has yet to set a standard across all cars and all locations. The problem is very much like the competition between Beta and VHS formats for videos. The standards were incompatible and proprietary, so the owners of the protocols were jockeying to dominate and reap the rewards. Portugal, Israel, and various regions have set policies to install recharging stations, but without a universal standard, it could be a poor investment.

High-speed recharging relates directly to the issue of vehicle range and its universal appeal. Families that can afford two cars might choose a conventional fuel vehicle for longer trips and a small EV for local commuting and errands. On the other hand, families or individuals buying a single car must choose one that serves multiple purposes. If electric cars have limited range and are unreliable, they will not sell. This is the fundamental theory behind the plug-in hybrid—one car can serve multiple purposes.

Peak Oil Demand

Conventional energy forecasts from oil companies, investment banks, the Energy Information Administration, the International Energy Agency, and others proceed by balancing forecasted demand and supply to determine price. The process requires the forecasters to analyze the size of the resource base, the number of new finds, field depletion, energy demand growth, and a host of additional underlying assumptions. Since there is no real certainty about many of these variables, particularly when looking decades into the future, the forecasts have missed most important turning points and have failed miserably to predict future oil prices.

Professor Adelman points out that the quantity of oil ultimately produced and consumed is "unknown and unknowable." In essence, peak oil *will not be determined by supply, but by demand.* At some point, oil will become too expensive for most of its present uses. The oil market will fade, in the same manner that kerosene replaced whale oil. This chapter focuses on key options to replace oil and the acceleration of innovation spurred by high prices. Oil is not only replaceable, but alternatives abound. This may be a disappointment to OPEC's member states or to peak-oil advocates and other Malthusians, but it is good news for everyone else.

The market has already spoken and substitutes are in process. The automobile manufacturing industry is investing heavily in electric vehicles, and for the first time, the economics, infrastructure, and technology are favorable. If battery costs drop as anticipated, EVs are about to break through the marketing and production barriers that have held them back. They are initially costly and there are bound to be technical glitches in their development, but they are no longer a dream; instead, a viable alternative to the use of diesel and gasoline transportation fuels is at hand.

One way to calculate expected long-term oil prices is to compare the present value of substituting alternative energy sources to the present value of future oil production. So, how does the cost of buying and running an EV compare to conventional fuels? Oil prices may have to be higher than the crossover point to provoke a transition, but ultimately costs will equalize.

Tesla provides sufficient data to do a comparison to oil prices. According to the company, the battery pack costs around $20,000. The car is similar in size and shape to the Lotus Elise, which averages about 24 mpg. So assuming the cost differential to move from gasoline to electric drive is the battery pack, gasoline prices are going to have to be very high to make the switch economic. Assuming 100,000 miles over eight years, at a discount rate of 6%, with an electricity cost at 16¢ per kWh from a mix of renewables, gasoline prices would have to average $8.17 per gallon to make the switch pay out.

A gasoline price (excluding road taxes) of $8.17 per gallon implies a crude oil price of about $325 per barrel. Using these numbers, oil is worth even more than many OPEC producers have claimed. This, however, is the high end of the market—the Tesla Roadster is selling for many reasons other than to avoid buying gasoline. If an EV costs $7,500 more than a gasoline car (the cost differential likely in a few years once production settles down) and power costs 8¢ per kWh, then the gasoline price equivalent for 100,000 miles is $3.23 per gallon, implying crude oil prices of $113 per barrel.

Crude oil producers, however, need to use present value discounting too, (recall the discussion of Hotelling's theory of depleting resources in chapter 3). If consumers choose electric vehicles, they are lost as gasoline consumers for eight or so years, until the car is decommissioned. The present value to a producer of a lost sale eight years hence is $75 instead of $113 per barrel. Assuming no production constraints, oil producers have an incentive to lower oil prices in order to prevent car buyers from going electric.

There remains substantial uncertainty about the cost differential between electric and conventional vehicles. Thus, the calculations just offered need constant revision as technology improves. Moreover, until consumers are confident about the reliability of the technology, and clear about the cost savings, there will not be a stampede to alternative vehicles. Likewise, rational oil producers have great incentive to postpone that shift by stabilizing oil prices at levels below alternative costs. Whether or not rationality will prevail in the oil market remains an unanswered question, and the path that oil prices might take to long-term equilibrium could indeed be rocky.

Rationality ought to prevail among oil consuming nations too; policy makers must allow the price mechanism to work. Even a brief review of the transition from coal to oil reveals an astonishing number of complex and intricate substitutions over the decades. Adam Smith's invisible hand fermented that evolution and it is hard to imagine that even the wisest policy maker in the nineteenth century could have foretold the outcome.

The Curse and Blessing of Oil

There are several key points following from a review of alternatives to oil and the search for a backstop technology. First, at this point in technological development, it appears that the primary substitute for oil will be electricity. This is due mainly to recent breakthroughs in battery technology, which promises electricity storage at a reasonable cost. As a liquid and energy-dense hydrocarbon, oil is well suited to serve the transportation market. It is, however, a depleting resource, and synthetic oils are likely to be costly with large energy penalties and environmental side effects. Moreover, the internal combustion engine is not very efficient for propulsion when compared to electric motors. In contrast, a variety of competitive resources can generate electricity, many of which are renewable and have minimum side effects.

During the Great Depression, Franklin D. Roosevelt sought to spread electricity to rural, as well as urban, dwellers. His administration set up the Rural Electric Authority and sponsored federal hydroelectric projects around the nation. After electricity had been successfully distributed to farmers throughout its service territory, the Bonneville Power Administration (the federal agency in the Pacific Northwest) surveyed its customers and asked their opinion about the benefits and change brought by electricity. The answers were surprising, with water pumping cited as the greatest benefit, not electrical lighting or refrigeration. Releasing humans from daily drudgery and backbreaking labor is the task of modern energy systems, and it is far too easy to lose sight of this crucial role.

Since the end of World War II, oil has been fundamental to the unprecedented increase in global prosperity. It is hard to imagine how coal or any other known energy resource could have built the modern world. Oil has been a major facilitator of international trade, and the economic growth that has accompanied its development has fostered remarkable human mobility, across the globe and in every strata of society. If William Jevons was right, if low-cost energy and mechanization are the enablers of economic growth, then oil is at the center of the story. The century-long creation and redistribution of wealth in the oil age has unlocked extraordinary intellectual capacity. More people are better educated than ever before and poised to create alternatives to the fuel that made it possible. Oil may be a depleting resource, but its key role may simply be that of a building block. Unlike the fallen cedars of Lebanon, the depleted resources of Easter Island, and other ruins of over-extended civilizations, the end of the oil age will not be the end of the modern economy, only the keystone in its construction.

Notes

1. Jevons is credited, along with Leon Walras and Carl Menger, with the development of the marginal utility concept as applied to demand theory.
2. Following the collapse of NeXT, Steve Jobs returned to Apple with some of his top programmers. Under his stewardship the Mac operating system evolved adopting many of the NeXT software innovations, without abandoning its customer base. It seems that even Steve Jobs learned the lesson.
3. Present plants for the PHEV allow charging at home with a 220- or 110-volt outlet in combination with available fuels such as gasoline or diesel.
4. The precise energy loss associated with synthetic fuels varies enormously depending on the technology. In the case of corn ethanol, some analysts claim (see chapter 2) that the total energy to produce the fuel exceeds its energy content. In the case of oil sands, it can take as much as 2.6 million Btu of gas to produce and reprocess a barrel of synthetic diesel. This is a 44% energy

loss compared to about 18% for conventional crude oil production and processing of petroleum products. Since there is substantial variation, table 5.2 assumes an additional 20% loss for synthetic fuels.

5. Vehicle battery charging represents new load thrust on the grid. As long as new power stations are from renewable or gas-fired sources, this calculation holds.
6. There are often large differences between MSRP and actual dealer prices, depending on supply and consumer demand. The purpose of this comparison is to demonstrate that if better efficiency results in higher car prices the benefits of reduced fuel expenditures may not offset the higher initial cost. At the time the OTA completed its study, the authors assumed gas prices would be around $1.50 per gallon.

ACRONYMS

AER	Annual Energy Review (DOE)
ANS	Alaska North Slope
API	American Petroleum Institute
APIG	API Gravity
APPI	Asia Petroleum Product Price Index
Aramco	Arabian American Company
BG	British Gas
BIS	Bank of International Settlements
BNOC	British National Oil Company
BOE	barrels of oil equivalent
BP	British Petroleum
Btu	British thermal units
CARB	California Air Resources Board
CBOT	Chicago Board of Trade
CEO	Chief Executive Officer
CFD	Contracts for Differences
CFTC	Commodity Futures Trading Commission
CNG	compressed natural gas
CNPC	China National Petroleum Company
CO_2	carbon dioxide
CT	combustion turbine
DB	Deutsche Bank
DC	direct current
DNR	Department of Natural Resources
DSMR	Distributed Steam Methane Reformation
E&P	exploration and production
EC	European Community
EEI	Edison Electric Institute
EFP	Exchange Futures for Physicals
EIA	Energy Information Administration
EIS	Environmental Impact Statement
EISA	Energy Independence and Security Act
EPRI	Electric Power Research Institute

ERA	Energy Regulatory Administration
EU	European Union
EV	electric vehicle
FERC	Federal Energy Regulatory Commission
FIA	Futures Industry Association
FPC	Federal Power Commission
FSU	Former Soviet Union
FTC	Federal Trade Commission
GATT	General Agreement on Tariffs and Trade
GM	General Motors
GSE	Government Sponsored Entities
GTL	gas to liquids
ICE	Intercontinental Exchange
ICP	Indonesian Crude Price
IEA	International Energy Agency
IOC	International Oil Company
IPE	International Petroleum Exchange
ISDA	International Swaps and Derivatives Association
JNOC	Japan National Oil Company
kg	Kilogram
L	Liter
Li-Ion	Lithium ion
LNG	liquid natural gas
LOP	Law of One Price
LPG	liquefied petroleum gas
MEG	Maghreb-Europe gas pipeline
MIT	Massachusetts Institute of Technology
MJ	Megajule
MMS	Mineral Management Service
mpg	Miles per gallon
MTBE	Methyl Tertiary Butyl Ether
MW	Megawatt
MWh	Megawatt-hour
NAS	National Academy of Science
NBP	National Balancing Point
NEB	National Energy Board
NiMH	Nickel Metal Hydrite
NOC	National Oil Company
NYMEX	New York Mercantile Exchange
OCS	Offshore Coastal Survey
OECD	Organization for Economic Cooperation and Development
OFF	Office of Fossil Fuels (DOE)

OPEC	Organization of Petroleum Exporting Countries
OSP	official sales price
OTC	over-the-counter
PADD	Petroleum Administration Defense District
PC	Personal computer
PHEV	plug-in hybrid electric vehicle
PIW	Petroleum Intelligence Weekly
PSA	Production Sharing Agreements
PUC	Public Utility Commission
SAGD	steam assisted gravity drainage
SEC	Securities and Exchange Commission
SG	specific gravity
SINOPEC	China Petroleum and Chemical Corporation
SPR	Strategic Petroleum Reserve
TAPS	Trans-Alaska Pipeline System
TEOR	Thermally Enhanced Oil Recovery
THUMS	Texaco Humble Union Mobil Shell
TOCOM	Tokyo Commodities Exchange
TPA	third-party access
WTI	West Texas Intermediate
ZEV	zero-emission vehicle

BIBLIOGRAPHY

1. Adelman, M.A. (1972). *The World Petroleum Market*, John Hopkins University Press.
2. Adelman, M.A. (1995). *The Genie out of the Bottle, World Oil since 1970*, The MIT Press.
3. Alaska Department of Natural Resources (2007). *Alaska Oil and Gas Report*, 2000 and 2007.
4. Argus (2008). *Argus Methodology and Specifications Guide*, December 2008, www.argusmediagroup.com.
5. Barclays Capital, Japan Equity Research (2009). "Electric Car Market Starting to Move," Barclays, May 26, 2009.
6. Becker, P. (1981). "The role of Synthetic Fuel in World War II Germany," *Air University Review*, July-August 1981.
7. Black, D. G., (1986). "Success and Failure of Futures Contracts: Theory and Empirical Evidence," Solomon Brothers Center for the Study of Financial Institutions, NYU, Monograph Series in Finance and Economics: Monograph 1986-1.
8. Bradner, T. (2005). "Oil lease sale lacks big bids of the state's past," *Alaska Journal of Commerce*, April 10, 2005.
9. British Petroleum BP (2009). *BP Statistical Review of World Energy* 2008, BP Web site.
10. Campbell, C.J. (1997). *The Coming Oil Crisis*, Multi-Science Publishing Company and Petroconsultants S.A.
11. Carlton, D. (1984). "Futures Markets: Their Purpose, Their History, Their Growth, Their Successes and Failures," The Journal of Futures Markets, vol. 1, no. 3, pp 237-71.
12. CFTC (2009). CFTC Web site.
13. Coase, R. H. (1937). "The nature of the firm," *Economica*, 4: 386-404.
14. Deffeyes, K. (2001). *Hubbert's Peak*, Princeton, NJ: Princeton University Press.

15. DeMott, J. (1982). "Exxon shuts down its Colony Project," *Time*, May 17, 1982.
16. Deutsche Bank (2008). *Oil and Gas for Beginners, A Guide to the Oil and Gas Industry*, January 7, 2008.
17. Deutsche Bank (2008). "Electric Cars, Plugged In, Batteries Must Be Included," June 9, 2008.
18. Eberhard, M. & Tarpenning, M (2006). "The 21st Century Electric Car," July 19, 2006, Tesla Motors Web site.
19. Fallon, W. (1995). "A Market is Born," *Managing Energy Price Risk*, Risk Publications.
20. Frankel, P. H. (1969). *Essentials of Petroleum*, Frank Cass & Co. Ltd.
21. General Accounting Office GAO (1980). *The United States Exerts Limited Influence on the International Crude Oil Spot Market*, Report to the Congress by the Comptroller General.
22. George, R. (2008). "Alberta's Energy Future, Focus on Oil," www.neb.gc.ca, February 8, 2008.
23. Houthakker, H. S. (1959) "The Scope and Limits of Futures Trading," in Abramovitz, Moses, et al., The Allocation of Economic Resources, Stanford: Stanford University Press, 134-59.
24. Hull, J.C. (2009). *Options, Futures, and other Derivative Securities*, Seventh Edition, Englewood, NJ: Prentice-Hall.
25. Jaffe, A.M. & Soligo, R. (2007) "The International Oil Companies," Baker Institute, Rice University.
26. Jevons, W.S. (1856). *The Coal Question*, Library of Economics and Liberty,
27. Kalhammer, Kopyf, Swan, Roan, and Walsh, (2007) *Status and Prospects for Zero Emission Vehicle Technology*, Report of the ARB Independent Expert Panel 2007, California Air Resources Board, April 13, 2007.
28. Kaufman, R. (2003). "Does OPEC Matter? An Econometric Analysis of Oil Prices," *The Energy Journal*, 2003.
29. Klein, B.; Crawford, R.A.; Alchian, A.A. (1978). "Vertical Integration, Appropriable Rents, and the Competitive Contracting Process," *Journal of Law and Economics*, vol. 21: 297-326.
30. Lee, R. (1998). *What is an Exchange?* Oxford University Press.
31. Lynch, M. (2003). "The New Pessimism about Petroleum Resources: Debunking the Hubbert Model and Hubbert Modelers," *Mineral and Energy*, 2003.

32. Mabro, R. (1975-76). "OPEC after the Oil Revolution," *Millennium*, London School of Economics Journal of International Studies, 191-199.
33. Mirant (2009). Mirant Web site.
34. National Energy Board (2007). *Canada's Energy Future, Reference Case and Scenarios to 2030*, November 2007.
35. Nehring, R. (1978). *Giant Oil Fields and World Oil Resource*, Rand Corporation, June 1978.
36. O'Hara, M. (1995). *Market Microstructure Theory*, Cambridge, MA: Blackwell Publishers, Inc.
37. OPEC (2008). *World Oil Outlook 2008*, OPEC Secretariat.
38. *Petroleum News* (2008, 1). February 24, 2008, p. 17.
39. *Petroleum News* (2008, 2). August 28, 2008.
40. Pimentel, D. and Patzek, T. (2005). Energy Input Ratios, Natural Resources Research (Vol. 14:1, 65-76).
41. Platts (2008). "Platts Crude Oil Methodology and Specifications Guide," December 2008, Platts website.
42. Prius History (2009). "Toyota Prius History," www.cleangreencar.co.nz.
43. Ripple, R. D. (2008). "Futures Trading: What is Excessive," *Oil and Gas Journal*, June 8, 2008.
44. Sampson, A. (1975). *The Seven Sisters*, Bantam Books – Viking Press.
45. Searchinger, T, et al. (2008). "Use of U.S. Croplands for Biofuels Increases Greenhouse Gases Through Emissions from Land-Use Change," *Science* 319.
46. Sultzberger, C. (2004). "An Early Road Warrior, Electric Vehicles in the Early Years of the Automobile," IEEE History, www.ieee.org.
47. Schumpeter, J. (1942). *Capitalism, Socialism, and Democracy.*
48. Simmons, M. (2000) "The World's Giant Oil Fields,"
49. Simmons, M. (2005). *Twilight in the Desert: The Coming Saudi Oil Shock and the Coming Economy*, Hoboken, NJ: John Wiley & Sons.
50. Smithsonian (2008). "EV1 electric automobile," America on the Move Collection.
51. Stern, J. P. (1998). *Competition and Liberalization in European Gas Markets*, The Royal Institute of International Affairs.
52. Stern, Roger (2006). "Oil market power and United States national security," January 2006, PNAS.

53. Shell Web site (2009).
54. Tabuchi, H. (2009) "Why Japan's Cellphones Haven't Gone Global," *New York Times* July 29, 2009.
55. Taylor, J., Van Doran, P. (2005). "The Case against the Strategic Petroleum Reserve," CATO Institute, November 2005.
56. Texas Comptroller (2008). "Window on state government, Chapter 28, government financial subsidies," www.window.state.tx.us.
57. *Time* (1935). "Oil to Market," March 13, 1935, www.time.com.
58. *Time* (2007). "1997 GM EV1," www.time.com, September 3, 2007.
59. Treat, J. E. (2000). *Energy Futures Trading Opportunities*, Tulsa, OK: PennWell Books.
60. Tussing, A. R. and Tippee, B. (1995). *The Natural Gas Industry*, second edition, Tulsa, OK: PennWell Books.
61. Van Vactor, S.A. and Tussing, A. (1987). "Retrospective on Oil Prices," *Contemporary Policy Issues*, Volume V, Number 3, pp. 1-19.
62. Van Vactor (2004). "Flipping the Switch, The Transformation of Energy Markets," www.econ.com
63. Verleger, Philip K., (1987). "The Evolution of Oil as a Commodity," *Energy Markets and Regulation*, edited by R.L. Gordon, H.D. Jacoby, and M.B. Zimmerman, MIT Press.
64. Williamson, O. E. (1985). *The Economic Institutions of Capitalism*, The Free Press.
65. Wolkoff, N. (2000). Executive Vice President, NYMEX: Testimony before the New York Senate, February 1, 2000.
66. Yergin, D. (1991). *The Prize, The Epic Quest for Oil, Money, and Power*, New York, NY: Simon & Schuster.

INDEX

D

E

F

G

H

I

J

K

L

M

N

O

P

Q

R

S

T

U

V

W–X

Y

Z